AF552545

Food Security and Irrigation

FOOD SECURITY AND IRRIGATION

By

Dr. M. Lakshmi Narasaiah

M.A., Ph.D.,
Professor and Head
Department of Economics
Sri Krishnadevaraya University
Post-graduate Centre,
Kurnool— 518 002
Andhra Pradesh

DISCOVERY PUBLISHING HOUSE
NEW DELHI-110002

First Published-2003
Reprinted - 2014

ISBN 978-81-7141-667-7

Published by

DISCOVERY PUBLISHING HOUSE
4831/24, Ansari Road, Prahlad Street,
Darya Ganj, New Delhi-110002 (India)
Phone: 3279245 • Fax: 91-11-3253475
E-mail:dphtemp@indiatimes.com

Printed at:
Infinity Imaging Systems
Delhi

PREFACE

Combating hunger and poverty is the central point of Bread for the world's mandate. In our view, that is not so much about the quantity of food produced in the world. On the one hand it's about its fair distribution and, on the other, the access of poor people to chances of jobs. Put another way, it's to do with access to purchasing power. In the case of agriculture, that is bound up with the question of how food is produced. Whether the technologies applied maximize employment or replace work with capital.

Hunger Through Surplus

The question of production, employment and distribution are tied closely to the general conditions for development. It is certainly not exclusively external economic conditions which account for hunger and under-development. Structural deficits, political conditions and wrong policies in Third World countries have become increasingly clear. However, it can still be noted that global economic framework conditions remain enormously important for the development of agriculture in the Third World.

Twenty years ago, Bread for the World publicly expounded the thesis 'Hunger Through Surplus' and had to take much criticism for it—above all from agro-economists. But since then the contradiction between the ever-growing mountains of agricultural surpluses in the northern hemisphere and the increasing dependence on food imports of the South has become ever more apparent. Out of 120 poor developing countries, 107 today are net importers of food.

The North's surpluses of dairy products, grain, beef and sugar—which because of their production costs are exorbitantly expensive—thrust their way on to world market and destroy local supply systems (which are cheap because of subsidies), regional trade flows, and the sales possibilities of potential Third World agro-exporters. Thus, the surpluses contribute to the situation that in many developing countries a policy of neglecting local agriculture can be continued with impurity.

Initially, the promise to work on the yawning gap between hunger and surplus in the world was upfront on WTO agenda. But the pattern of explanation was well simplified. It said that surpluses arose only in those countries which supported their agriculture positively and, in fact, partly excessively. And that agricultural deficiencies in countries of the South were caused mainly by deprivation of resources and capital. However, the concept of not only reducing neglect of agriculture in the South but also its oversubsidising in the North to a sensible degree and thereby eliminating their distortions of world markets had a great intellectual attraction. At any rate, it promised more justice in agriculture.

Dr. M. Lakshmi Narasaiah

Contents

1

Food Security: Availability and Access to Food

The world food situation has never been better. Enough food is being produced today that, if it were evenly distributed, no one should have to go hungry. World food production is increasing faster than population growth: per capita production increased by 5 per cent during the 1980s. Real food prices are at historic lows and have been declining for some time now. Yields of major cereals have more than doubled in the past three decades. These trends have contributed to complacency in some quarters regarding the world food situation.

Yet, more than 700 million people in the developing world do not have access to sufficient food to lead healthy and productive lives. More than 180 million children are underweight. Diseases of hunger and malnutrition are widespread. The desire to satisfy food needs has, in combination with increasing population densities and inadequate agricultural intensification, led to much degradation of environmentally fragile lands, such as forests and steep hillsides.

Over the next 20-30 years, farmers and policy makers in developing countries will be challenged to provide food at affordable prices for almost 100 million more people every year—the largest annual population increase in history. Moreover, they will have to increase food production from more productive use of the land and without further degradation of natural resources: area expansion is no longer a feasible option in most of the world.

What future food security will look like depends not on exogenous factors over which we have no control but on the decisions and actions taken by the major players: households, private-and public-sector agencies, governments, and the international community. If we continue to act as we have in the 1980s and early 1990s, more people will suffer from food insecurity, it will be because some or all of these players failed to act in an appropriate and timely manner.

Feeding the World: Availability and Access to Food

There is enough food in the world today to feed everyone, if it were evenly distributed. Availability of daily food energy per capita in the developing countries as a whole increased by 0.7 per cent per year during the 1980s.

Twenty-five developing countries, including about half of the African countries, were unable to assure sufficient food energy (2,200 calories per person per day) for their populations at the end of the 1980s even if available food energy were evenly distributed within each country. This is down from 45 countries at the end of the 1970s.

However, available food is neither evenly distributed nor fully consumed. Availability of enough food at global, regional, or natural levels does not necessarily mean that everyone is well fed. For people to be food secure—that is, to have access at all times, to the food required for a healthy and productive life—there must be both availability of food and access to food. Access to food by households (and individuals) is conditioned by poverty: the poor usually lack adequate means to secure access to food.

Over 1.1 billion people in developing countries were living in poverty in 1993, more than 500 million in conditions of extreme poverty. South Asia is the home of about 50 per cent of the developing world's poor—more than 500 million people. Another 15 per cent are found in East Asia, 19 per cent in Sub-Saharan Africa, and 10 per cent in Latin America and the Caribbean. The prevalence of poverty (the proportion of each region's population that is poor) is very high about 50 per cent—in South Asia as well as in Sub-Saharan Africa.

Today, there are more than 700 people who do not have access to sufficient food to meet their needs for a healthy and productive life; they often go hungry adults and children also suffer from diseases associated with hunger and poverty. For almost one fifth of the total population of developing countries to be chronically hungry tarnishes the image of a world that is now considered food-secure because it produces enough food.

Great progress has been made in meeting food needs during the last 30 years. For instance, the number of underfed people declined from an estimated 976 million in 1974-76 to 786 million in late 1980s. But the problem is far from solved. Keeping up with increasing needs and demands due to population growth, income increases, and dietary changes is itself a formidable challenge.

Hunger and food insecurity have a significant effect on health and nutrition of both adults and children. They can lead to growth failure in children. About 184 million pre school children in developing countries were underweight in 1994. About 55 per cent of these underweight children were found in South Asia and another 16 per cent in Sub-Saharan Africa. The proportion of children that the underweight is higher in South Asia (almost 60 per cent), but it is also significant in Sub-Saharan Africa (30 per cent) and Southeast Asia (31 per cent). It is worrysome that the number of underweight children in Sub-Saharan Africa during the 1980s from 20 million to 28 million is particularly striking.

In addition to energy deficiencies, micro nutrient deficiencies are also widespread in the developing world. About 14 million pre school children (under the age of five years) have eye damage as a result of Vitamin A deficiency. Ten million of these children are found in Southeast Asia. Between 250,000 and 500,000 preschool children go blind each year due to Vitamin A deficiency, two-thirds of these children die within months of going blind. Many more children are mildly affected. Recent research has shown that even mild deficiencies can increase mortality significantly. Vitamin A deficiencies are closely linked to diet, which can be influenced by agricultural research and policy.

Iron deficiency affects about 1 billion people in the world, particularly children and women of reproductive age. Iron deficiency leads to anaemia, which if not checked can diminish learning capacity and increase morbidity and morality. In the developing countries, about 370 million women between 15 and 49 years of age—42 per cent of this population group—where anaemic in the 1980s. Almost one-half were in South Asia. And there are tentative indications from South Asia and Sub-Saharan Africa that the prevalence of anaemia is rising in non-pregnant adult women of reproductive ages.

In Sub-Saharan Africa, this trend is undoubtedly associated with deterioration in general standards of living, including increased poverty and food insecurity. Anaemia partly arises from diets insufficient in iron, which again could be addressed through agricultural research and policy. For example, a possible reason why iron deficiency and anaemia are going up in South Asia may lie in the decrease in production of iron rich pulses during that same period, which in part reflects the larger research input into competing crops such as wheat in South Asia. This emphasizes the importance of considering the effects on diet and thus on health and nutrition in setting research priorities for yield-increasing research.

South Asia is the home of about half of the developing world's hungry and food—insecure people, but this population group is growing rapidly in Sub-Saharan Africa. Much of the poverty and food insecurity is in rural areas, mainly in low—potential areas such as arid zones, but urban poverty is also growing rapidly.

Four Key Factors will Influence Future Food Production and Consumption

Global and regional food production and consumption during the next 10-20 years will be influenced by a large number of factors. Changes in the following four sets of factors are likely to particularly important:

1. Economic growth and economic policies;
2. Population growth and urbanization;
3. Rural infrastructure, agricultural production technology, and access to modern inputs; and
4. Natural resource management and environmental considerations.

The expected impact of each of these factors on future food production and consumption is considerable.

Economic Growth and Economic Policies

Economic growth must resume in the developing world, especially in Sub-Saharan Africa. To support such growth, it is critical to

- complete structural adjustment and economic reforms;
- remove external barriers to growth such as trade distortions and subsidies in developed countries;
- liberalize trade and remove market distortions;
- enhance access by the poor to land, capital and technology;
- expand investment in rural infrastructure, health, education, and agricultural research and technology;
- facilitate sustain ability in agricultural production; and
- reverse the decline in international assistance to agriculture.

Growth in real per capita income during the 1980s was disappointing for developing countries as a whole. However, the low average rate of growth covers large variations among regions. The high rates of economic growth in Asia are expected to continue through the 1990s, while incomes in Sub-Saharan Africa are expected to keep pace with population growth.

Future economic growth depends or internal policies as well as on the international policies as well as on the international environment. The extent to which current structural adjustment and economic reforms in Latin America, Sub-Saharan Africa, the Commonwealth of Independent States (CIS), Eastern Europe, and selected countries in Asia and the Middle East are carried to successful completion at an appropriate speed and sequence is a paramount importance for future economic growth in those countries.

Closely related to this issue is the question of the most appropriate role of the state in a market-oriented economy with inappropriate institutions, poor infrastructure, and insufficient experience by the private sector in dealing effectively in a competitive market environment. Overreaction to past failures such as excessive and inappropriate state intervention may cause governments to take on a passive role where intervention is needed to assure that the markets function effectively and to deal with outside influences on the economy.

Future economic growth will also depend on the international trade environment, including trade distortions by developed countries, and access to external aid. Import restrictions for agricultural and non-agricultural products in Japan, the European Union, and the United States, along with domestic agricultural subsidies and implicit and explicit export subsidies for agricultural products, are of particular concern.

Population Growth and Urbanization

If progress in economic growth is not to be undermined by rapid population growth and excessive urbanization, effective population and migration policies are necessary to complement growth-oriented policies. Such policies must focus on:

- universal access to family planning information and technology; and
- incentives to reduce rural-urban migration, such as provision of employment in rural areas and stimulation of agricultural and non-agricultural growth in rural areas.

Although the annual growth rate is falling for the world as a whole, the population increase during the next 20-30 years, of slightly less than 100 million people a year, will be the largest ever. Approximately 97 per cent of this increase is projected to occur in the Third World, with Africa alone accounting for 34 per cent of the growth. Thus although reductions in annual population growth rates have begun to occur in Asia and Latin America, they are insufficient to counter the absolute increases. Population growth rates of these magnitudes will greatly increase the need for food and other basic necessities.

Rural Infrastructure, Agricultural Production Technology, and Access to Modern Inputs

Continued progress in all three of these areas is critical to future food security:

- Resources must be committed to infrastructure construction and maintenance. Labor-intensive public works programmes are a viable mechanism for building roads, reforesting areas, and engaging in soil conservation projects, while creating employment and income in rural areas.
- International and national agricultural research must continue to develop yield—enhancing production technology, especially in maize, millet, and other crops, as well as build tolerance or resistance in crops to pests and adverse climatic conditions.
- Farmer access to modern inputs must be facilitated through provision of credit and technical assistance. Inputs must be made available to all farmers on time and in required amounts.

The importance of investments in rural infrastructure within the context of rapid urbanization has already established. Even without rapid urban growth, however, such investments are needed in many developing countries, particularly the poorest ones, to facilitate agricultural and rural development. Improved rural infrastructure enhances access to export markets, modern production inputs, and consumer goods. It reduces marketing costs, promotes exchanges, between intracountry markets, reduces spatial and temporal price distortions, and, in general, increases efficiency in production and marketing.

However, while essential, effective rural infrastructure alone is not enough to assure agricultural and rural development and rapid increases in food production in developing countries. Yield enhancing production technology is of critical importance. Although opportunities for expansion of agricultural production into lands not currently under cultivation still exist in some countries, such opportunities are so limited that they would probably not be able to counter losses of current agricultural lands to alternative uses on a global level. Furthermore, attempts to expand agricultural production into new lands would, in most cases, require large investments in technology, tools and materials and would increase the risk of land degradation and deforestation. Thus, future increases in food production must come primarily from higher yields per unit of land rather than from land expansion.

Agricultural research has successfully developed yield-enhancing technology for the majority of crops grown in temperate zones and for several crops grown in tropical zones. The dramatic impact of agricultural research and modern technology on wheat and rice yields in Asia and Latin American since the mid-1980s is well known. Less dramatic but significant yield gains have been obtained from research and technological change in other crops, particularly maize.

Natural Resource Management and Environmental Considerations

Research, technology development, incentives, and regulations are needed to prevent environmental degradation.

These measures include appropriate water management policies, reduction of subsidies that encourage wasteful use of inputs, better definition of ownership and user rights to resources including land, education of farmers to encourage appropriate use of technology and resource conservation, and the provision of alternatives to resource-degrading inputs and techniques. Since poverty is a major source of degradation, poverty eradication is justified also on environmental grounds.

The recent surge in public and private concerns about negative environmental effects of economic growth and development may, if sustained, have important implications for agricultural development and future food production and consumption. Of particular concern of the need to avoid degradation of natural resources such as land and water, as well as deforestation, water contamination, and health risks associated with the use of chemicals. Since most of the current and potential resource degradation and environmental contamination result from situations in which those who cause and possibly benefit from degradation do not pay the costs, neither the market nor the individual producers and consumers are likely to incorporate preventive measures into their behaviour. Only when sufficient damage has been done to influence significantly current or future production costs will market and producer behavior change. The state is more likely to undertake preventive measures either through publicly funded research and technology development or through incentive policies and regulations. Extensive water logging, salination, and associated land degradation and productivity losses resulting from inappropriate water management are of particular concern in large parts of Asia.

No Time for Complacency

Population growth will outstrip growth in food production in Sub-Saharan Africa for a long time to come unless more is done to accelerate agricultural growth. In the year 2000, the population growth was more than 3 per cent a year, while food production was 2 per cent or less a year. In 2000, the production shortfall was increased to about 50 million tons of grain. The region will not have the necessary foreign exchange to import such large amounts of food. And African governments will not be able to count on enough food aid to make up the difference. If current trends continue, by the year 2020, Africa will have a food shortage of 250 million tons, which is more than 20 times the current food gap.

Poverty is expected to increase rapidly in the coming years. Sub-Saharan Africa's share of the world's poor was 28 per cent in 2000. Furthermore, the number of underweight children is expected to increase in Sub-Saharan Africa.

Asian demand for cereals was at an annual rate of 2.1 per cent in the year 2000, where as food production was 1.9 per cent. Much of the production shortfall is likely to be dealt with through expanded imports and perhaps through expanded regional production in response to price increases.

In Latin America, by contrast, growth in food production is anticipated to exceed food demand growth; food production was to grown by 3 per cent annually between 1990 and 2000, while food demand to grow by 2.5 per cent per year.

Large areas of land are rapidly being degraded and deforested. And the principal reasons for environmental degradation—poverty, high population growth, and limited access to appropriate agricultural technology—are not being dealt with effectively.

About 700 million people are food insecure for them the food crisis has arrived. For the 10-12 million pre-school children who died in 1994 from hunger and diseases related to malnutrition, the food crisis came and went. One-third of the pre-school children of the Third World are unable to grow to their full potential and face increased risk of death and disease.

Complacency is not in order. Clearly, Malthus underestimated the power of science to expand food production. The mass starvation that was predicted for Asia in the 1970s and 1980s did not occur because science was effectively put to work to expand crop yields. However, past yield increase came about people with foresight made appropriate decisions. The failure to expand investments in agricultural research and technology development during the 1980s and 1990s indicates that such foresight no longer prevails. Given the long lag time between investment in agricultural research and the resulting production increases, failure to invest today will show up in production shortfalls 10 to 20 years from now. The problems associated with environmental degradation will present themselves sooner. We must not wait until a global food crisis is upon us or until the last tree has fallen to make these investments.

2

India's Food Challenge

Is India's population growing disproportionately to its food supply? Will famine once again hit millions of people? Most agriculture experts agree that a Malthusian crisis is not likely to occur in the near term. The reason, the overall food situation in India has been characterized by a large increase in regional output since the famine-ravaged 1960s.

At that time, the food situation was described as 'desperate' in India. Famine had plagued India's Bihar state in the sixties. International food specialists predicted further famine because food production looked as if it would lag far behind population growth. Instead, average crop yields per acre soared, thanks to the introduction of high-yielding varieties of rice and wheat and to expand irrigation and chemical fertilizer use. It has been called the 'green revolution'.

Double Role of Irrigation

The keys to the higher food production have been irrigation, the adoption of high-yielding varieties (HYVs) of foodgrains and the increased use of modern inputs such as fertilizers. Irrigation has played a double role, it has not only helped raise yields through synergistic interaction with HYVs and fertilizers, but has also contributed to considerable increases in harvested area by enabling higher cropping intensity.

Still, there are ominous clouds on Indian food horizon. In light of the region's high population growth, increased urban sprawl and rampant environmental degradation, there are signs that hunger problems could loom unless action is taken by Indian's and international development agencies.

Shrinking Base

The favourable food supply situation is likely to disappear within the next decade, due to a shrinking resource base. The earlier decades had witnessed a natural resources based growth strategy as there was adequate land and water resources for development. But this is fast disappearing due to urbanization, industrialization and ecological degradation. We should also remember that about 50 per cent of food production is from rainfed lands and a few years of drought could alter the food security which we now enjoy. The high costs of irrigation and land development, coupled with low commodity prices, are also hampering required investments and these effects will be seen in the next decade.

"By the year 2030, India will have to produce 60 per cent more rice with much fewer resources. Clearly, there will be a major challenge for scientists and policy-makers to meet the increased food demand. India's population is growing 2 per cent a year, making the challenges for regional food security a daunting task.

The solution for meeting future food demand will be breakthroughs in science and technology since yields levels have reached a plateau and are even showing signs of decline. The possibilities through biotechnology and genetic engineering are exciting and can herald another 'green revolution'. This is the only hope for avoiding the Malthusian dilemma.

In gauging the region's population-food squeeze, it is useful to look first at its swelling population. India—the world's second populous region contains several states with high population growth rates.

Ironic Problem

Rapid population growth dilutes and impedes economic development. An increase in the population base puts greater pressure on finite resources, both financial and natural, and, in the context, worsening of income distribution, increased poverty incidence and environmental degradation.

Moreover, there is the ironic problem, that although rapid population growth increases poverty, poverty encourages larger families through its impact on access to education and decreased prospects for child survival.

If population growth is uncontrolled, the economic and social consequences are:

- ecological imbalance, with greater pressure on natural resources;
- increased urban crowding, with increase in demand for municipal services and infrastructure;
- a more unequal income distribution, particularly as labour supply outpaces job creation;
- signs of mass poverty, including high infant and child mortality rates, high levels of child malnutrition and hunger, poor school performance, unemployment and underemployment.

Bleak Prospects

Existing population growth rates is unsustainable, even for the relatively near future. Unless population growth rate is kept within manageable limits, the prospects for creating acceptable standards for living for low-income groups in India will be bleak.

The Indian population is growing more rapidly than ever before and will continue to do so for at least four decades. Indeed, without major technological breakthroughs and changes in patterns of consumption, even the most optimistic population growth projections are likely to be accompanied by increase in poverty, hunger and environmental degradation.

Whether we look at population, the environment or development, the next 10 years will be critical for our future. The decisions we make or don't take will widen or narrow our options for a century to come. They could decide the fate of the earth as a home for human beings.

3

Food Production

During the last 25 years, world agriculture successfully expanded food production faster than population growth. This can continue for the next 25 years and beyond, if appropriate action is taken. Although world food stocks are currently low and grain prices high, the world is not about to run out of food. We can produce enough food for future generations if we choose to do so.

The widespread food insecurity, unhealthy living conditions, and abject and absolute poverty in many developing countries are already threatening global stability. Failure to assure sustainable food security will foster the very conditions that will further destabilize and polarize the world in the years to come with tremendous consequences for all people.

The Basic Facts

Poverty is widespread in developing countries, with over 1.1 billion people living on a dollar a day or less per person. Human resource development in developing countries is lagging: 1 billion people lack access to health services, 1.3 billion do not have access to adequate sanitation systems, and one-third of primary school enrolls drop out by Grade 4. Natural resources, upon which future food production depends, are being degraded at alarming rates: almost 2 billion hectares of land have been degraded in the past 50 years: about 180 million hectares of forests have been converted to other uses during the 1980s, marine fisheries are collapsing around the world, and regional and seasonal water shortage afflict many developing countries. Improved appropriate technology is essential to increase productivity. Yet low-income food deficit developing countries are grossly underinvesting in agricultural research and many are reducing their support.

It calls for sustained action in *six* priority areas. *First*, we must selectively strengthen the capacity of developing country governments to perform appropriate functions such as establishing or clarifying property rights, promoting private-sector competition in agricultural markets, and maintaining appropriate macro economic environments. Predictability, transparency and continuity in policy making and enforcement must be pursued.

Investing in People

Second, we must invest more in poor people in order to enhance their productivity, health, and nutrition. It is not only unethical but economically wasteful that a large share of the world's population is malnourished, illiterate, sick, and without access to productive resources. Access to primary education, primary health care, reproductive care and family planning information, and clean water and sanitation must be assured for all people. Access by the poor to productive resources and remunerative employment must be improved. Empowerment of women must be supported.

Third, we must accelerate agricultural productivity. Agriculture is the life blood of the economy in low-income developing countries. In those countries, it provides upto three-quarters of all employment and half of all incomes. There are very strong links between agricultural productivity increases and broad-based economic growth in the rest of the economy. Agriculture is an engine of growth in low-income developing countries. National and international agricultural research systems must be mobilized to develop improved technologies focused on developing countries, and extension systems must be strengthened to disseminate the improved technologies and techniques. Low-income countries currently spend less than 0.5 per cent of the value of agricultural production on agricultural research compared to 2 per cent spent on agricultural research in middle and high-income countries. An increase of agricultural research expenditures in low-income countries to at least 1 per cent of the value of a agricultural output is urgently needed, with a longer term target of 2 per cent. National agricultural research must be supported by a vibrant international agricultural research system that undertakes research with large international benefits applicable across boundaries. Current investments in international agricultural research are grossly inadequate to provide the support needed by developing countries. It is of critical importance that agricultural research result in reduced unit-costs of production. Such cost reductions will make food economically accessible to low-income consumers, and permit producer incomes to increase. To assure relevance of research and appropriate distribution of responsibilities, interactions between public sector agricultural research system, farmers, private enterprises, and NGOs must be strengthened.

Fourth, we must assure sustainability in agricultural production and sound management of natural resources. Farmers, local communities, and governments must be encouraged to establish and enforce systems of rights to use and manage natural resources, to improve the way water is allocated and used, to reverse land degradation where it has occurred, to reduce the use of chemical pesticides and promote integrated pest management programmes, and to implement integrated soil fertility programmes in areas with low soil fertility. Local control over natural resources must be strengthened and local capacity for organisation and management improved. Investments in less-favoured geographical areas, that is, areas with agricultural potential, irregular rainfall patterns, and fragile soils must be expanded. Most poor people in developing countries reside in rural areas, and most rural poor reside in less-favoured areas. Yet, most investments, including agricultural research investments, still focus on the more favoured areas. If we are serious about reducing poverty and protecting the natural resource base, the balance between the less-favoured and more-favoured areas must be redressed.

Fifth, we must reduce food marketing costs in low-income developing countries. The cost of bringing food from the producer to the consumer is very high in many of these countries. Efficient, effective, and low-cost agricultural markets must be developed in order to bring these costs down. Inefficient state-run firms in agricultural in-put markets must be phased out; investment in developing and maintaining infrastructure, especially in rural areas, must be forthcoming; policies and institutions that favour large-scale, capital-intensive market agents over small-scale credit and savings institutions must be facilitated; and technical assistance to create or strengthen small-scale, labour-intensive competitive rural enterprises must be provided.

Sixth, we must expand and realign international development assistance. Many years ago, industrialized countries had agreed to allocate at least 0.7 per cent of the Gross National Product (GNP) to international assistance. Most countries have not reached or do not maintain this target. Not only must the industrialized countries increase international development assistance to reach the 0.7 per cent target, but they must realign it to low-income developing countries. Also contrary to the middle and higher-income developing countries, the poorest countries are not able to gain access to capital from the rapidly expanding international commercial capital market. Developing countries in turn must seek measures to diversify sources of external funding, stem capital flight; and improve the effectiveness of the aid they receive.

4

Food First

By the time this day is over, about 40,000 human beings—mostly children—will have died from hunger, malnutrition and related causes. Today and every day the deaths will mount, reaching an annual toll of 13 to 18 million. Few of these people will have been caught up in famine or other emergencies. Most will have suffered from a 'silent' assault—the kind that seldom makes the headlines, but which claims its victims just as relentlessly.

It is intolerable that such deprivation and suffering should be allowed to exist in a world of potential food plenty. Having enough food is fundamental to all else. At the most basic level, this may entail humanitarian relief to assist people in emergency situations. In the transition from relief to development, however, we must look at systems for ensuring that societies have the capacity to produce or purchase the food they need and that it is accessible to all.

Sustainable food security, fuses the goals of household food security and sustainable agriculture; it requires both. It requires looking not only at the aggregate supply of food, but also at the distribution of income and land, and at other issues: Do people have enough income to buy food? Enough land to grow their own food? Does the food distribution system deliver food where it is needed? How much food is wasted due to inadequate distribution systems? What are the implications of trends in population growth for future food needs? What is the status of women in society, and what opportunities do women have to alter rapid population growth rates? What is being done to regenerate the resource base for food production? These questions need to be asked and answered in every country.

The challenge of sustainable food security is immense, and it is growing. One billion people—20 per cent of the global population—are too poor to obtain enough food to sustain normal work. Half a billion are too poor to obtain the food needed for healthy growth of children and minimal activity of adults. Today's failure to feed people, however, may be but a prologue to a much larger failure in the future. Given likely population increases, world food output must triple over the next 50 years if the world's people are to have a nutritionally adequate diet. It will be difficult enough to achieve this expansion under favourable circumstances, and conditions may be far from favourable.

For example, according to recent estimates an area of about 1.2 billion hectares—the size of China and India combined—has experienced moderate to extreme soil deterioration since World War II as a result of human activities. Over three-fourths of that deterioration has occurred in the developing regions from causes such as overgrazing, deforestation, land clearing, unsound agricultural practices and increased soil salinity and water logging, largely from irrigation. Other environmental threats to the agricultural resource base include loss of water and genetic resources, adverse effects of pesticides and climate change, both local and global.

At the most aggregate level, the required increase in food production could be met if production grew at the historic average, that is, at the two per cent per annum rate achieved over the past half-century. But is this realistic? To produce three times more calories, all the land currently under cultivation around the world would, within 50 years, have to attain levels of productivity as high as those exhibited by the very best cropland today.

To this challenge add the possibility of diminished returns from the technological, energy and other inputs that have made agriculture so successful. Some experts believe that most of the potential for increased output of cereals—from improved plant varieties, from increased use of pesticides and fertilizers and from expanding the area under irrigation—has already been captured.

Viewed from this perspective, the goal of achieving sustainable food security in the decades ahead emerges as one of the greatest challenges humanity has ever faced. Agricultural output must be tripled, and people, must have the income to buy the food they need. The erosion of the resource base must be halted and then reversed. Failure on any of these fronts will yield unprecedented human suffering.

What will it take to achieve sustainable food security? Obviously, the effort will have to be immense, both in size and complexity. Outlined below are a few simple (but no easy) steps that are absolutely essential elements of serious effort.

First, as citizens of the world, we must all come to see sustainable food security as a fundamental aspect of global peace and human security. This goes well beyond merely denouncing the use of food as a weapon.

Second, we must adopt concrete international goals, such as reducing world hunger by half over the next 10 years. We will never achieve the goal of sustainable food security unless we aim at specific milestones, and assess rigorously our progress in moving toward them.

Third, we must forge a true global partnership, a compact for sustainable food security. All countries—rich and poor—have important roles and responsibilities. There must be reciprocal responsibilities among nations, not one-way transfers.

Fourth, we must see deterioration of the agricultural resource base—terrestrial, aquatic and climatic—for what it is: a major threat to development and a major source of economic loss. Farmers are the largest group of environmental decision—makers in the world. We must ensure that they have the means to make sustainable development a reality where it counts—in the fields and fisheries.

Fifth, we must empower the people who work the land and who keep it productive. They are in the best position to decide the most appropriate ways to graft new technology onto their own traditional knowledge of seed selection, plant protection and nutrient-cycling. Special emphasis should be given to the role of women, the main providers for two-thirds of the poorest households in the developing world, as well as the producers of 60 per cent of all food grown consumed locally.

Sixth, we must build the capacities of developing countries, both the government and in civil society. Capacity-building means empowerment for self-reliance. It means strengthening national capacities, both inside and outside government. This is essential for recognition and analysis of problems, for decision-making on courses of action and for management of systems and processes.

Seventh, not only must we build capacity in developing countries, we must also create linkages among researchers in industrial and developing countries. This will help minimize the time lag between discovery and practical utilization. In addition, analysts from various countries must work together to examine future food security issues with different scenarios of population growth, agricultural productivity, markets and trade, climate change, loss of soil and bio-diversity and, last but not least, political instability, in order to devise options for rational choices.

We know a good deal about how to rid the world of the scourge of hunger, and how to begin to move toward sustainable food security on a global basis. We know that economic growth and prosperity are necessary, though not sufficient, conditions for eradicating hunger. We also know that development efforts must encompass not only food production, but also socio-economic factors, including sustainable livelihoods for poor families, the implications of population growth rates, the status of women and girls and so forth. We also know that good words are not enough. No more than ever before it is crucial that we marshal the political will to achieve our goals.

5

Food for the Billions

Will there be enough food to feed 8 billion people who will live on earth in 25 years' time? Surprisingly few people, at least in the industrial countries, seems to be overly concerned with this question. Whereas the world conferences on the environment, on women, human rights or social issues which were held in recent years were preceded and accompanied by intensive public debate, food does not seem to be a burning issue. Don't we have mountains of surplus food, people ask. Do we not have to pay our farmers to leave their land idle in order not to add to the glut on the world markets? And hasn't the Green Revolution ended famine even in countries like India which used to be a synonym for hungry people? So where is the problem?

The advance made in agricultural production since beginning against a background of imminent crisis are indeed remarkable. In only 20 years, yields of major crops like rice, maize and wheat in developing countries went up by 80 per cent, outpacing even the rapid increase in population. But this growth in yields has slowed down in recent years, and the aim of 'food for all' is once again becoming elusive. About 800 million people still do not have access to enough food to meet their basic daily needs, nearly 200 million children suffer from protein and energy deficiencies, 88 countries—44 of them in Africa—have a deficit in food production.

Every one wants to increase food security. The definition is that "food be available at all times, that all persons have means of access to it, that it be nutritionally adequate in terms of quantity, quality and variety, and that it be acceptable within the given culture". To achieve this goal, more food must be produced—much more, because we must not only adequately feed the 5.8 billion people already on earth, but also the additional two billion who will be added to world population in the next 25 years. Critics argue that the problem is not one of production alone, but one of poverty elimination. People are not hungry because there is no food, but because they have no money to buy, it these critics say. Available resources must be better distributed to end hunger in the world.

However, even if we succeed to eliminate poverty in the next few decades—a feat which appears highly unlikely—there would still be the need to boost production, because with rising incomes people also want to eat more and better food including meat. As can already be observed in the countries of East Asia, the newly acquired wealth leads to higher consumption levels which puts additional strains on available resources are getting scarcer. Agricultural lands are being degraded at alarming speed by erosion, salinity, desertification or disappear altogether due to urban or infrastructure development. It has been estimated that 40 per cent of productive land now has diminished capacity to supply benefits to humanity due to direct human impacts of land use. Water for agricultural purposes is getting scarcer almost everywhere, and there are hardly any land reserves to be brought into production to widen the agricultural base.

In this situation, there is no alternative to increasing and improving production from the existing land area. This can only be done through research which finds the best varieties which will bring the highest yields at the lowest cost to the environment. Sustainable agriculture is the key notion—one that maintains biodiversity, uses as little chemical inputs as possible and does not over exploit water and soil resources.

In recent years, agricultural research has been neglected—partly because of the erroneous belief that with mountains of meat and lakes of milk further production increases were not desirable. Since global grain production has stagnated and world stocks have reached an alarmingly low level last year, there has been a noticeable change of mind. To raise the awareness among governments around the world that promotion of agriculture is urgent if hunger is to be avoided in the next century.

Important work is already being done by the international agricultural research institutes which promoted the Green Revolution in the sixties and seventies and are now again in the forefront of finding solutions to the daunting task of feeding 8 billion people by the year 2020. The International Rice Research Institute (IRRI) in the Philippines, the Maize and Wheat Research Institute (CIMMYT) in Mexico or institutes like ICARDA in Syria and ICRISAT in India which work on agriculture in semi-arid and dry areas, are all seeking solutions to the problem of raising production while at the same time preserving the environment. These institutions as well as national agricultural research institutions need all the support from the public and, of course, appropriate funding, to help them accomplish their task.

The scientists are optimistic that they can develop the varieties and farming systems which will allow mankind to feed everyone on earth well into the next century. But the task is not for the scientists alone. An economic and political order must also be in place which makes it possible to eradicate poverty and allow everyone to enjoy the benefits that science can offer. Feeding the billions is, therefore not only a scientific, but first and foremost a political.

6

Less Food Security in the South?

Combating hunger and poverty is the central point of Bread for the world's mandate. In our view, that is not so much about the quantity of food produced in the world. On the one hand, it's about its fair distribution and, on the other, the access of poor people to chances of jobs. Put another way, it's to do with access to purchasing power. In the case of agriculture, that is bound up with the question of how food is produced. Whether the technologies applied maximize employment or replace work with capital.

Hunger through Surplus

The question of production, employment and distribution are tied closely to the general conditions for development. It is certainly not exclusively external economic conditions which account for hunger and under-development. Structural deficits, political conditions and wrong policies in Third World countries have become increasingly clear. However, it can still be noted that global economic framework conditions remain enormously important for the development of agriculture in the Third World.

Twenty years ago, Bread for the World publicly expounded the thesis 'Hunger through surplus' and had to take much criticism for it—above all from agro-economists. But since then the contradiction between the ever-growing mountains of agricultural surpluses in the northern hemisphere and the increasing dependence on food imports of the South has become ever more apparent. Out of 120 poor developing countries, 107 today are net importers of food.

The North's surpluses of dairy products, grain, beef and sugar—which because of their production costs are exorbitantly expensive—thrust their way on to world market and destroy local supply systems (which are cheap because of subsidies), regional trade flows, and the sales possibilities of potential Third World agro-exporters. Thus, the surpluses contribute to the situation that in many developing countries a policy of neglecting local agriculture can be continued with impurity.

Initially, the promise to work on the yawning gap between hunger and surplus in the world was upfront on WTO agenda. But the pattern of explanation was well simplified. It said that surpluses arose only in those countries which supported their agriculture positively and, in fact, partly excessively. And that agricultural deficiencies in countries of the South were caused mainly by deprivation of resources and capital. However, the concept of not only reducing neglect of agriculture in the South but also its oversubsidising in the North to a sensible degree and thereby eliminating their distortions of world markets had a great intellectual attraction. At any rate, it promised more justice in agriculture.

Subsidies Can Make Sense

To avoid misunderstandings, we have nothing against the support of agriculture in Europe. Above all not when it is done for social, ecological or agriculturally beneficial reasons.

On the contrary, agriculture's important role for food security, the sustainable handling of natural resources, the settlement of rural areas, and the social function of family farms justify a special position for it is economic life, including protection and support.

But that must not be carried so far that surpluses are produced with the help of dubious production methods and then dumped on the world market at markedly less than cost price, causing incalculable damage in the poor countries. On the other hand, purposeful, promotion of rural development is a prerequisite and model for greater self-sufficiency worldwide, especially in Third World countries.

Complementary Functions of World Markets

The poor counties of the South have no alternative than to become self-sufficient in food. The World markets can at best assume complementary functions. The countries would take indeterminable risks if they integrated themselves completely in the world markets, and thereby wanted to make themselves dependent upon global agro-markets. These are and will remain extremely unreliable factors that are conditioned by enormous fluctuations in prices and quantities, the powerful, and in many cases obscure, influences of multinational concerns, the manifold political interventions in the agricultural scene in most countries, and the dangers of social and ecological dumping.

But when we now look at the results of the WTO, we are disappointed. The development question and the balancing of hunger and surplus are finally no longer on the agenda. Programmes to increase food production in the poor countries were not the priority of the negotiations. The liberalization of agro-policies in the developing countries would have meant making the disadvantaging of their farmers the subject of international negotiations. That did not happen.

On the contrary, the concepts developed with an eye on the reform of agricultural policy in the North, which target the reduction of the support level, are to be transferred to the South without questions. To be sure, there are a whole number of exemptions for the poorest developing countries. But the WTO results have also set the trend there, namely the dismantlement of subsidies. We cannot understand how such a thing can be demanded as a policy programme, especially for Africa. Support for African agriculture is largely absent, *i.e.* there is absolutely nothing to dismantle. That's why many international conferences repeatedly emphasize the need for these countries to achieve a greater degree of self-sufficiency in food by stronger support of their agriculture.

Agro-Dumping

Certainly, some changes have been made in the North's agro policy system which will also have positive impacts on world agricultural markets. However, also here we must express our disappointment. Agricultural dumping will continue. The only difference will be the new policy instrument of direct transfer of income instead of subsidised grain prices. The opening of markets in future will hardly go beyond the current preference conditions.

The entire set of W.T.O. agreements, however, bears the imprint of the two agricultural superpowers, the USA and the European Union, which make mutual concessions and coordinate their agricultural policies. But one hears nothing about the target of freeing the world agricultural market from unnecessary distortions and ensuring justice. The intention of the agro-superpowers was solely to defend their global market shares.

The development aid agencies cannot close their eyes to these problems. On the contrary, in future they must make very much greater efforts in suggesting better goals, programmes and instruments which are capable of forming a policy that can then be included in the agenda of the next rounds of negotiations. We may perhaps have slept a bit through the past WTO talks. Therefore it is even more important that we get very much more involved from now on.

7

Genetic Diversity and Food Security

Maintaining a diversity of crops and varieties is a key to survival for millions of farmers living on impoverished land. For thousands of years, farmers have used the genetic variation in wild and cultivated plants to develop their crops and raise new breeds of live stock. Genetic diversity gives species the ability to adapt to changing environments, including new pests and diseases and new climatic conditions. Plant genetic resources—that component of genetic diversity of actual or potential use to humanity—provide the raw material for breeding new varieties of crops. These, in turn, provide a basis for more productive and resilient production systems that are better able to cope with such stresses as drought or overgrazing and can reduce the potential for soil erosion. The use of genetic diversity—on-farm, through field experimentation or in sophisticated gene transfer procedures—remains arguably the best route so securing our food and that of our children.

Although science has made enormous strides in improving the world's ability to feed itself over the past three decades, we cannot afford to rest idle. Nearly 800 million people in the developing world do not have enough to eat. In these regions, the rural poor represent about 73 per cent of the people living in poverty. They often live in marginal or unsuitable farming areas, such as zones with saline soils, and conditions, or degraded or hilly areas. Often isolated from other farms and far from urban areas, many poor farmers have barely benefited from agricultural developments elsewhere. In many cases they do not have access to commercially bred high yielding crop varieties. Diversity flourishes and remains important under such conditions.

Selections and Breeding

Poor farmers are well aware of the relationship between the stability and sustainability of crops and crop varieties on their lands. Their management and use of a diverse range of plants has often helped them to survive under the most difficult conditions. By growing a range of different crops, farmers have a better chance of meeting their needs. These might be crops that mature at different times or that can be easily stored to help to ensure a stable food supply throughout the year. They may also help farmers provide a nutritionally balance diet for their families, exploit different environment niches that exist on their land, or diversity their income sources.

Importantly, the genetic diversity contained in different varieties provides farmers with options to develop, through selection and breeding, new and more productive crops that are resistant to pests and diseases. The result may be a vast range of local varieties of crops grown by farmers in any one area.

Not respecting diversity can incur high costs: in 18th century Ireland, where potatoes were the only significant source of food for about one third of the population, farmers came to rely almost entirely on one very fertile and productive variety, which proved susceptible to the devastating potato blight fungus. The resulting famine caused the death or emigration of more than 20 per cent of the population.

The value of diversity goes well beyond its ability to support stable production systems in marginal environments. As the world's human population rises, environmental problems (desertification, deforestation, erosion etc.) are intensifying, climate change, particularly global warming, could bring about drastic changes in the location of the world's agro-ecological zones. Farmers will require new crop varieties capable of producing under diverse conditions, without adding ever-increasing amounts of fertilizers and other agro-chemicals. Because of the limited scope for growth in the world's cultivated areas, each new generation of varieties will have to be more productive than its predecessors.

Much has been written about the use of genetic engineering in plant breeding. Modern molecular techniques can be used to transfer genes from one living organism to another or to change the genetic material within to produce more desirable traits. Genetic Engineering has enormous potential to help solve problems that have proved intractable using conventional breeding approaches, such as developing crop varieties with in-built resistance to key pests and diseases and tolerance to stresses such as drought. However, the possible impact of these techniques, particularly on human health and the environment, is giving rise to fierce world wide debate.

Take the case of banana and its close relative plantain, two of the developing world's most important crops. Their improvement is hindered by the sterility of most cultivars, a problem that can be addressed through genetic engineering. It is now possible to transfer gene constructs, such as those associated with disease resistance, directly into varieties with other desirable characteristics, drastically reducing the need for pesticides.

Today, research on genetic engineering is focussed on the development of commercial varieties of the world's major crops of interest in industrialized farmers. Many of the staple crops of importance to poor farmers in developing countries, such as cassava, bananas, beans and yams, have received relatively little attention. This situation is likely to continue as plant breeding is increasingly privatized and biotechnology becomes the fast-growing province of private industry. Meanwhile, the high costs of the new technologies are quickly exceeding the capacity of many, if not most, public research institutions—both in developing and developed countries—to support them. Thus, for the time being, increasing agriculture's role in the development of the world's poor is likely to continue to depend on the identification, maintenance and use of genetic diversity'.

8

Grain Production

The relationship between the growth in world population and the grain harvest has shifted over the last half-century, neatly dividing this period into two distinct eras. From 1950 to 1984, growth and the grain harvest easily exceeded that of population, raising the harvest per person from 247 kilograms to 342, a gain of 38 per cent. During the 14 years since then, growth in the grain harvest has fallen behind that of population, dropping output per person from its historic high in 1984 to an estimated 317 kilograms in 1998—a decline of 7 per cent, or 0.5 per cent a year.

These global trends conceal widely divergent developments among countries, contrasts that can be seen for the world's two most populous nations: India and China. In both, grain production per person was close to 200 kilograms as recently as 1978. Since then, the figure in India has edged up slightly but still falls short of 200 kilograms, while in China production has surged since the economic reforms in 1978, with per-person output now at nearly 300 kilograms. The combination of a dramatic surge in grain production and an equally dramatic reduction in population growth has given China a large margin of safety, effectively eliminating most of its hunger and malnutrition. Meanwhile, although India has also achieved impressive gains in its harvest, these have been largely cancelled by population growth, leaving its 976 million people living close to the margin.

What has happened in China and India is the story of developing countries in general. The overwhelming majority have achieved substantial, if not dramatic, gains in their grain harvests over the last half-century. Some, such as Thailand, have combined this with a much slower growth of population, which means that agricultural gains translate into rising grain production per person. In Pakistan, by contrast, grain production per person climbed steadily for a while, but it peaked in 1981 at 186 kilograms. Since then it has been declining nearly 1 per cent a year. In effect, Pakistan's farmers are losing the battle with population growth.

The slower growth in the world grain harvest since 1984 is due to the lack of new land and to slower growth in irrigation and fertilizer use. Irrigated area per person, after expanding by 4 per cent since then as growth in the irrigated area has fallen behind that of population.

The increase in world fertilizer use has slowed dramatically since 1990, as diminishing returns to the application of additional fertilizer has stabilized use in the United States, Western Europe, and Japan and slowed annual growth in world fertilizer use from 6 per cent between 1950 and 1990 to scarcely 2 per cent in recent years.

Although Malthus was primarily concerned with the additional demand for grain generated by population growth, rising affluence is also playing a role. In a low income country such as India, grain consumption per person is less than 200 kilograms per year and diets are typically dominated by a single starchy staple-rice, for instance. With scarcely a pound of grain available a day per person, nearly all must be consumed directly, leaving little for conversion into animal protein. For the average American, on the other hand, the great bulk of the 800-kilogram daily grain consumption is taken in indirectly in the form of beef, pork, poultry, eggs, milk, cheese, ice cream, and yogurt. At the intermediate level, in a country like Italy, people consume 400 kilograms of grain a day. Future food price stability thus depends on expanding production fast enough to keep up with both population growth and rising affluence.

One question often asked is, How many people can the Earth support? This must be answered with another question, At what level of consumption? If the world grain harvest of 1.87 billion tons were expanded to 2 billion tons in the years ahead, it would support 10 billion Indians or 2.5 billion Americans. To answer the question of how many people the Earth can support, we first have to know the level of consumption we expect to live at.

Now that the frontiers of agricultural settlement have disappeared, future growth in grain production must come almost entirely from raising land productivity. Unfortunately, this is becoming more difficult. After rising at 2.1 per cent a year from 1950 to 1990, the annual increase in rainland productivity dropped to scarcely 1 per cent from 1990 to 1997. The challenge for the world's farmers is to reverse this decline at a time when cropland area per person is shrinking, the amount of irrigation water per person is dropping, and the crop yield response to additional fertilizer use is falling.

9

Cropland

Since mid-century, global population has grown much faster than the cropland area. The trend is likely to continue in the next century, dropping cropland per person to historically low levels. The ever smaller per capita cropland base will make food self-sufficiency impossible for many countries, and will test the capacity of international markets to meet a growing demand for imported food.

For millennia, farmers satisfied rising food demand by bringing new land under the plow. But by mid-century cropland expansion could no longer meet the food needs of an increasingly populous and prosperous world. The 10,000 year era of steady expansion was over, and a new era began that stressed raising land productivity. As this high-yielding era shows signs of faltering, concern over the shrinking supply of cropland per person looms ever larger.

Since mid-century, grain area—which serves as a proxy for cropland in general—has increased by some 19 per cent, but global population has grown 132 per cent, seven times faster. Largely as a result, grain area per person has fallen by half since

1950, from 0.24 to 0.12 hectares. Assuming that grain area remains constant, grain area per person will fall to 0.07 hectares by 2050. In crowded industrial countries such as Japan, Taiwan, and South Korea, grain area per capita today is smaller than the area of a tennis court.

As grain area per person falls, more and more nations risk losing the capacity to feed themselves. Having already seen per capita grain area shrink by 40-50 per cent between 1960 and 1998, Pakistan, Nigeria, Ethiopia, and Iran can expect a further 60-70 per cent loss by 2050—a conservative projection that assumes no further losses of agricultural land. The result will be four countries with a combined population of more than 1 billion whose grain area per person will be only 300-600 square meters, less than a quarter of the area in 1950.

The historical record suggests that such a small area per person will send a substantial share of a country's people to world markets from their food. Consider the experience of six countries in East Asia whose per capita grain area currently ranges from 200 to 600 square meters per person. Sri Lanka relies on imports for more than a third of its grain, while Japan, Thiwan, South Korea, and Malaysia buy more than 70 per cent of their grain from abroad. North Korea is the only one of the six that does not import heavily (it gets less than 20 per cent of its grain requirements from abroad), but its population is poorly fed-indeed, on the verge of starvation.

The concern is that population growth will push many nations—not just the four fastest-growing ones—below the 600-square meter-threshold in coming decades. In Asia alone, where grain area per poised to Cross the threshold person stands at 800 square meters 16 countries are by 2050, and many of them much sooner. As this process unfolds, the number of people who will turn to foreign markets for their food will likely jump sharply. These countries will find an increasingly tight international grain market, with nations from the Middle East, North Africa, and other regions already buying a third or more of their grain overseas.

In addition to per capita losses, population growth can lead to degradation of cropland, reducing its productivity or even eliminating it from production. As a country's population density increases and good farmland becomes scarce, poor farmers are forced onto ecologically vulnerable land such as hillsides and tropical forests. In the Philippines, for example, hillside agriculture accounted for only 10 per cent of all agricultural land in 1960, but 30 per cent in 1987. Because it is highly erodible, hillside land is easily damaged; worldwide, some 160 million hectares of hillside farmland—11 per cent of cropland—were characterized in 1989 as 'severely eroded.' Similarly, population pressure can force peasants to overfarm the poor soils of tropical forests. After being cleared and farmed for a few years, these soils typically require fallow periods of 20-25 years, but population pressures keep poor farmers on the same land for far longer than the soil can support, cutting fallow periods to just a few years in some areas of tropical Africa and Asia.

Finally, population pressures on a fixed base of land can result in rural landlessness. In Bangladesh, for example, landlessness among rural households rose from 35 per cent in 1960 to 53 per cent in the early 1990s. Interestingly, Balgladesh is regarded as a success in slowing population expansion, as its growth rate declined from 2.8 per cent in the late 1970s to 1.5 per cent in the early 1990s. But its success came too late to prevent the increase in rural landlessness, highlighting the need to work sooner, rather than later, for population stabilization.

10

Health Care Relief in Conflict Situations

What Can We Learn from the Food Relief Experience?

Conflicts and war occur in many of the poorest nations where populations already suffer from severe ill-health. War leads to an increase in disease and to a worsening of the already fragile condition of populations. Health care itself becomes a victim of conflict. Many deaths which occur during these emergencies are not discretely related to the conflict itself but are the result of lacking access to public health services. Furthermore, conflict itself but are the result of lacking access to public health services. Furthermore, conflict contributes to the deterioration of already pre-existing structural weakneses of the health care system. An example is the period of internal conflict in Uganda (1970-1986) when health services declined in the aftermath of the war due to the impact of foreign assistance and the planning vacuum in which the activities took place.

The Impact of Conflict on Health Care

Conflict and civil strife may lead to a major disruption of health services. This is not only a result of physical destruction but also of finding shortages since national governments increase spending on military activities. Casualties increase the demand for curative services which can divert already limited resources from preventive care.

In the case of the Sudanese civil war a large majority of health professionals was forced to abandon rural health professionals was forced to abandon rural health services and left for urban areas or neighbouring countries in order to find new employment. Entire preventive health services such as immunization as well as water and sanitation projects collapsed leaving the population exposed to infectious diseases and epidemics. In urban areas, the gap in public health care provision is sometimes filled with the expansion of private services. In rural areas, private sector involvement in health care is rather marginal, apart from some omission hospitals or pharmacies. Therefore the non-formal health care sector often makes a substantial contribution towards health care.

With the rise of internal conflicts in Africa, more people suffer from emergency situations. This also increases the influence and impact of international donors. External assistance nowadays accounts for more than 25 per cent of government health expenditure in sub-Saharan Africa.

The size of donor involvement reflects the power of international agencies to control the policy domaine. Countries in conflict or post-conflict situations are under pressure to 'rescue' their health systems and accept global policies in exchange for aid assistance and relief.

However, in the period after 1991, donor organisations tended to increase their expenditures for high profile humanitarian operations rather than ordinary development activities. This shift may reflect the increasing influence of media covering some of the conflicts. Too often, organisations intervene with ad-hoc assistance without sufficient consultation at local level.

Donors' Perceptions in Designing Relief Interventions

Today, in many parts of sub-Saharan Africa development assistance has virtually collapsed and has been substituted by relief assistance. The problem is that relief interventions are based on a Western construction of reality, reflecting what is desirable and necessary in times of conflict. Most interventions therefore stress physical and material needs, presuming that the social aspect of food and health is not an immediate issue to address.

The question which arises here is on who's views and perceptions these needs are based? While donors interests may be guided from the perspective of ill-health, the recipient government may be concerned with the collapse of the economy. However, any intervention needs to take into account that local knowledge and practices are shaped by state interests as well as power relationships. The common belief that health care systems always collapse due to conflict is sometimes mistaken. Considering the fact that today's internal conflicts are often fragmented, conflicts do not necessarily result in a breakdown of the health care delivery system.

Donors tend to respond with a 'package' approach and developing countries ministries of health increasingly play a symbolic role. The evidence suggests that international organisations tend to create vertical programmes which undermine national public health programmes. Foreign interventions are technically sophisticated and reorienting health are towards a more curative approach. Too little attentions given to strengthen the health care system within its own limits, providing more appropriate technology, drugs and emphasizing the training of local health staff.

Another vital issue concerns the existence of already fragile health information systems. Agencies tend to bring their own systems which leads to further fragmentation. The local perspective on what are the 'basic needs' in physical and social health are usually not considered. Health relief interventions do not recognize the potential of the communities and the non-formal health sector such as healers and traditional midwifes in supporting and maintaining health care sector presents a substantial contribution towards health. It is not the question between choosing either allopathic or traditional services, it is more the decision which kind of illness will be best treated by which practitioner. There is a need in further exploring the role of this sector particularly since this is sometimes the only service available for certain populations.

Responding to Local Needs

More community-based public health interventions could be vital to reduce mortality and morbidity. For example in Somalia during the 1992 war and famine high mortality rates due to measles and diarrhea could have been prevented by involving the communities in primary health care activities such as immunization and nutrition improvement.

In the African context Tigray is an example where health service had been sustained and partially expanded during the civil war against the Ethiopian government. Local government structures called baitos promoting social and economic development. Baitos encouraged communities to establish revolving funds for drugs and medical equipment. It actually functioned as an early type of community financing system.

As mentioned above, the challenge in changing health care relief strategies is to overcome the approach of short-term interventions, particularly in a changing conflict environment where conflicts are complex and interruptions are no longer short-term. Therefore interventions need to be linked with the process of conflict resolution to avoid health care or food aid being used by politically dominant groups.

Food Relief in Conflict Situations

Food interventions have both a survival and a production function. For example, food-for-work may be part of an income programme or food aid can be monetized to generate local currency. However, food aids has to be seen beyond the objective to fulfill nutritional goals, it also defines relationships between social groups in regard to food accessibility and how food is shared. Food aid is aiming to meet people's basic food requirements and minimising risk and severity of disease by complementing services such as basic health care.

In more stable political conditions where free food aid is given it presents an income transfer by releasing income which normally is spent on food. However, in conflict situations food relief frequently becomes part of the dynamics of conflict such in the case of Sudan where it is used to sustain the struggle between the North and the South without resolving it. Furthermore, the military attack food supplies in the fight against rebels who depend on the support from the communities.

Health is also a matter of food security. Where food insecurity coincides with conflict situations, health and survival are threatened. Food security provides some concepts on how and why vulnerable households manage to survive in period of hardship (coping strategies).

Coping Strategies in African Trouble Zones

Today, most conflicts in Africa such as the ones in the Great Lake Region, Angola or Congo cause major problems of food insecurity. They are linked to the civil wars which produce substantial social disruption as a result of massive population movements. The analysis of coping strategies showed that household respond to these conflict situations by eating less, selling livestock and land, or trying to find new sources of income.

In some emergency situations, however, such coping mechanisms may fail. In the case of the war in Mozambique food aid was vital since coping strategies were limited and people had to sell all their assets which was particular true for internationally displaced persons and refugees.

It has been argued that food relief bypasses local structures in favour of those qualifying on a nutrition status criterion, decided by international organisations, or it may attract populations to refugee camps to receive free food rations and thereby undermines local production. In the case of Rwanda food aid was targeted at the internally displaced and left out the local population. This can be due to donor bias in needs assessment.

Food scarcity is not always so result of civil war but its creation may be rather a political objective. An example is food manipulated by local elites and the military like in the case of Sudan. It can be summarized that generally relief operations often bear the risk of fueling the process of instability and violence rather that helping to contain the situation.

Lessons From Food Relief for the Health Sector?

Through the experience of food relief in recent civil wars such as Sudan, Somalia, Mozambique etc., there has been an increasing awareness of the economic and political context in which operations takes place. Like food relief, health care is a political tool which can, if not properly targeted, undermine peoples access to health care services. While food production is linked to food security, it is more difficult to identify factors leading to self-sufficiency in health care.

As mentioned above, food aid is aiming to insure survival. It also has an economic aspect, protecting household assets. Health care relief is targeted to assure immediate physical survival based on the importance of social health. Unfortunately, curative interventions hardly consider the socio-cultural dimension of health. Therefore it would be beneficial if health care interventions consider local norms and traditions. Interventions should be compatible and complement local health programmes. The emphasis should be on strengthening formal and non-formal health institutions both in service provision and training.

In food relief, distribution and needs assessment identification are controversial issues for discussion. While the programme design is shaped by donors perceptions, the actual programmes are influenced by the priorities of some powerful leaders as well as the socio-economic and political context.

Health care interventions need to analyse these issues in the context of economic and political systems in order to identify the most vulnerable groups, for example populations living in areas which are more, operations require a stronger involvement of communities both as users and active participants to carry out and maintain public health programmes.

There is a need for a new concept to be designed which applies to chronic emergencies. In the absence of a policy framework, guidelines need to be developed in order to overcome the inconsistency in planning and implementation. Donors need to change their assumptions on which they plan their health relief responses. A starting point in improving the efficiency of these operations is to provide institutional support to local authorities and organisations and involve them in the planning and implementation of programmes.

11

The Future of Agricultural Trade

In the Uruguay Round, countries recognised that the long term solution for agriculture did not lie in administered prices, trade restrictions, supply controls, and export subsidies but rather in open, non-distorted markets. It is the time to take bold steps toward bringing agricultural trade into the 21st century by accelerating agricultural trade reform.

There are four key areas for accelerating reforms: eliminating export subsidies; increasing market access though substantial tariff cuts and expansion of tariff-rate quotas; cutting further trade-distorting domestic subsidies; and ensuring technical standards are based on sound science.

The world's farmers and ranchers are facing two difficult challenges at the down of the 21st century. First, they are being asked to provide more products at lower cost, higher quality, greater variety, and in a safer manner than ever demanded before. Second, they are being asked to produce this abundance on a shrinking natural resources base that is often subject to government regulations. Meeting these global challenges will require unleashing the production potential of world agriculture while practising proper environmental stewardship. The ingenuity and hardwork we usually associate with farmers will be essential to meet these challenges, but they will not be sufficient unless we further reform agricultural, trade to create an environment that rewards risk and investment and encourages efficiencies.

Today's Agricultural Challenges

Farmers are responsible for feeding a rapidly growing world population. And despite progress over the years, too many people still are not getting enough food. Many countries including the United States, are working vigorously to promote technological innovations to meet the need for food and fiber in the coming years. However, as important as this work is, it is only part of the solution. These technologies and the hard work of the world's farmers need a trading environment that encourages investment and efficient production, and generates economic growth to finance production and consumption needs long-term trends in agriculture pose serious challenges for all farmers. The same technological advances that increase yields may result in lower prices. Increasing social concerns about effect of agricultural production on the environment and living conditions result in new restrictions on farm activities. As urban dwellers and industry stake competing claims for land, water, and energy, many producers find their ability to farm made ever more difficult.

Two approaches to organising the agricultural economy present a stark contrast in dealing with these challenges. One model, popular in Europe and Asia, is to retain an inward-looking agricultural system focused on supply control and government regulation geared to keeping farm prices high and, since guaranteed high prices are a drain on the treasury, to controlling production. Under this approach, bureaucrats try to assess the optimal level of national production—not so little that imports are needed and not so much that excess production; must be bought at high prices and; then dumped on world markets. This 'command-and-control' structure stifles farmer efficiency and ingenuity and distorts world markets, especially as subsidized surpluses are regularly exported; and it does not address the challenge to farmers to produce food for the next century. It also ignores the interest of domestic consumers (who have to pay high internal prices) and producers in other countries (who have to compete with subsidized products). Of biggest concern is that the anti-market policies of this approach hamstring the agriculture sector from pursuing the technological advances needed to meet its future challenges.

Another approach is to place agriculture on a more market-oriented basis, particularly by removing trade barriers and reducing trade-distorting policies. Greater market orientation was the principle that actions agreed to in the last set of multilateral trade negotiations. In the Uruguay Round, countries recognized that the long-term solution for agriculture did not lie in administered prices, trade restrictions, supply controls, and export subsidies but rather in open, non-distorted markets. Now is the time to take bold steps toward bringing agricultural trade into the 21st century by accelerating agricultural trade reform.

The Gains From Trade

The benefit from free and fair trading of agricultural products have immediate effects on people. Eliminating trade barriers and reducing unfair competition will help ensure that farmers have incentives to produce and consumers have access to the products they desire. Liberalizing agricultural trade will contribute to better resource allocation by farmers, which has conservation benefits, rewards low-cost producers, encourages efficiencies, and removes the drag on economic growth.

Opening trading opportunities also increases the food security of food-importing countries by giving supplier countries the confidence required to put more land into production and to create marketing relationships. Trade provides consumers with year-round access to a greater variety of less expensive products, which rewarding producers who are able to find and meet specific consumer demands for high-value products. In a broader context, by allowing imports that are more efficiently produced elsewhere, trade encourages specialization in efficient agricultural and non-agricultural production.

More dramatically, trade literally saves lives. Without the international flow of food products from area with abundant production to areas where food is scarce, many people in the world would be eating less or not at all. Trade has dynamic effects, as well, that push long-term productivity growth. For example, access to customers in overseas markets creates an incentive for technological innovation, resulting in exciting developments in improved seed varieties and production techniques. International markets also expand market outlets, raising prices and giving producers increased confidence to produce more than required merely for national needs, allowing productive farmers to not only feed neighbour but literally feed the world.

Equally important, trade in agricultural products is becoming increasingly critical to farm and ranch incomes. Increased productivity and often times flat domestic demand increases the importance of reliable international markets. Foreign markets are not just a dumping ground for surplus products; overseas consumers value choice and quality, particularly when producers in their own country cannot meet their demands or when they are charged inflated prices. Consequently foreign and value-added agricultural producers, raising farm-gate prices and helping support the range of agriculture-related industries.

Political reality also encourages a focus on international markets; policies based on high government guaranteed prices are ultimately political untenable because they are hugely expensive, unresponsive to the needs of customers and producers, incentive to environmental and agronomic realities, and the shameful waste of economic assets. Rather than farming government programmes, our producers are looking for customers around the world.

While agricultural trade benefits consumers and producers alike, it is an area in which progressive reform is ardently opposed by entrenched domestic interests. Producers in some countries, cosseted by high guaranteed prices and protective tariffs, oppose any move toward greater market orientation. Intervention in the agricultural economy—measured by the Organisation for Economic Cooperation and Development by summing price supports, direct payments, and other support as a per cent of total agricultural production—has actually increased in some countries from the levels at the beginning of the Uruguay Round. In the last set of multilateral trade negotiations, countries began the process of dismantling protection and delinking farm support from production decisions. Consequently, reforms have been undertaken by some countries.

The WTO Opportunity

The major objective in the upcoming farm talks is to accelerate the reform process initiated in the Uruguay Round. That means further substantial negotiations on tariffs, subsidies, and other trade-distorting measures so that the level and other trade-distorting measures so that the level and direction of trade are determined by market forces, not government intervention.

Four key areas are outlined below:

(i) *Export Competition*: Export subsidies are the most distorting trade tool because the level of direction of trade is directly determined by government subsidies. Today, the European Union (EU) is the only substantial export subsidizer—nearly all other countries agreed not to use, or have only limited recourse to use, export subsidies in the last round of negotiations. EU farmers, responding to domestic prices frequently twice the world price, produce more products than can be consumed in Europe, but at such high prices that they can be sold abroad only with generous subsidies. These subsidies push other competitive suppliers out of the market (which is expensive and unfair) and discourage production in countries that have a competitive advantage in agricultural production (which is wasteful and is threatening both to the environment and to future farm production needs).

In the Uruguay Round negotiations, countries acknowledged the corrosive nature of subsidies and agreed to cap and reduce their use. The upcoming negotiations should eliminate them to ensure that countries do not resort to other policy tools that allow government spending to determine winners in the marketplace. Specifically, WTO members should look closely at curbing distorting state trading agricultural export monopolies that can disguise subsidies and exert distorting market power, along with other policies used to dispose of surplus commodities on a non-market basis.

(ii) ***Market Access*:** Measures applied at the border to stop trade currently are the principal barrier to a freer and more open trading environment for agriculture. Market access barriers deny efficient producers the chance to compete in other markets and limit the variety and quality of products available to consumers. Opening markets and maximizing trade opportunities and fundamental principles of WTO, and we still have a long way to go in agriculture to open markets to competition.

The Uruguay Round Agreement set agricultural trade on a more predictable basis by requiring that all non-tariff measures, such as quotas and import bans, be converted to simple tariffs. While this was a necessary first step to removing trade barriers, many of the tariffs are still prohibitively high. For example, while the average tariff assessed by the United States on agricultural products is less than 5 per cent (and nearly zero for industrial products), the average agriculture tariff-rate quota (TRQ). Where only specific quantities of imports receive low duties. Many other commodities also; are subject to high tariffs.

As we start the next century, higher tariffs should not stop the flow of imported agricultural products. Where TRQs remain as a transitional step before we achieve more open trade, we expect more specific disciplines on the way in which they are administered. Similarly, we need to take a hard look at agricultural state trading monopoly. Importers; use of these state traders may have been justifiable when more restrictions allowed on farm trade, but in the tariff-only regime it is hard to see why a government needs to insert itself between an exports and an end-user.

(iii) ***Domestic Subsidies*:** Domestic subsidy programmes are often the root cause of other-distorting policies. Subsidy

policies that increase domestic prices above world price levels can be maintained only if price-competitive imports are restricted. Additionally, over production generated by high domestic prices can be sold on world markets only with export subsidies that bring the price down to the world price. While reining in distortive domestic subsidy programmes has value in its own right for rationalizing agricultural production, the WTO negotiations will focus on their trade-distorting elements.

In the Uruguay Round negotiations, countries, agreed to distinguish trade-distorting subsidies (generally those linked to the production of a specific crop or related to price supports) from non-trade distorting subsidies (such as research and development, training and environmental production). The trade-distorting subsidies were capped, and the process of reducing allowable levels of subsidies began. This distinction is a good one: the nasty sort of subsidy that distorts markets and strait jackets producers should be cut, while programmes that will increase a country's ability to produce agricultural products in the next century without distorting production incentives should not be reduced.

(iv) ***Standards***: As WTO members make progress on cutting tariffs and subsidies, the temptation increase to disguise trade barriers as health and safety measures or other innocuous-sounding 'technical standards'. Moreover, when regulations purportedly designed to protect health are instead vehicles for domestic protectionism, the credibility of the entire safety apparatus of a country is put up for questioning. When good science is replaced by politics, the basis for sound health policy is undermined. Therefore, increasing government accountability by putting the emphasis on sound science for health standards should discipline disguised barriers to trade and strengthen health policy.

In the Uruguay Round, countries agreed to a set of sound principles: each has the right to maintain health and safety measures, but these must be based on sound science, backed by scientific evidence and an assessment of the risk, and be no more trade-restrictive than required to meet health goals. In practice, countries have found that these principles work well—bogus measures adopted without scientific basis have been successfully challenged in the WTO without sacrificing health concerns. Creating a supportive environment for the propagation of yield-enhancing biotech products also is critical for meeting the needs of the coming century.

Agriculture is Different

Agriculture occupies a special place in the national economies of most countries around the world. Farmers are responsible for feeding and clothing people. Farming also holds a powerful claim on our national cultures that calls for the preservation of rural lifestyles and values. Farm production is subject to the cruel vagaries of weather and the relentless decline in prices and increases in costs. Some people point to these factors as justifying a different treatment for agriculture in the international economy, including justifying trade-distorting agricultural policies. This is wrong-headed; societies can support farms and preserve rural communities in ways that foster choice, protect natural resources, and expand trade.

Farm production in the next century cannot afford to be trapped in a static system in which prices are determined by government mandate, production decisions are controlled by central planners, and farmers are forced to produce only for local consumers. This myopic system cannot be sustained in any important agriculture producing society. Moreover, this type of system will not meet the needs of the coming century, when we will face unprecedented consumer demand and natural resource constraints.

Instead, I look forward to dynamic world of agricultural trade in which producers, exporters, and retailers apply the creativity of the human mind to the natural bounty of the earth. In this 'new' world, we will produce a greater amount and variety of food than ever before, feed the coming billions, sustain our environment, and unlock economic resources otherwise stifled by moribund protectionism, ultimately raising living standards around the world.

12

Developing Countries and the WTO Agricultural Negotiations

Developing countries as a group have much to gain from continued progress towards a transparent, rule-based trading system in agriculture. The researchers say that negotiations should eliminate export subsidies, impose stricter disciplines on export taxes, cut tariffs, and ensure that food aid continues to be available to poor countries in grant form and delivered so as not to displace domestic production in the countries receiving it. Badly managed food aid, or cheap food imports due to export subsidies, may just reinforce the bias of economic policies against the rural sector. With its negative impact on poor agricultural producers, 'they say. International research organisations (such as IFPRI, among other institutions) may provide support to developing countries through programmes of collaborative research, technical assistance and capacity strengthening.

Starting with the first round of trade negotiations under the General Agreement on Tariffs and Trade (GATT) after World War II, there has been a relatively steady trend of increasing multilateral trade liberalization. The successive rounds of negotiations recognized the greater needs of developing countries, especially since the Tokyo Round. Yet the participation of developing countries was limited. Since many developing countries were not members of GATT, the major forum for airing their views was provide by the United Nations Conference on Trade and Development. The views of developing countries had some impact on the Lome agreements and on aid flows, but had limited influence on negotiations concerning trading rules, which were discussed within the frame work on the GATT, where OECD (Organisation for Economic Cooperation and Development) countries set the agenda.

In the Uruguay Round, which began in 1986 and concluded in 1993, developing countries played a larger role in the negotiations compared to previous rounds. In particular, agricultural net exporters organised the Cairns Group (which in addition to Australia, New Zealand, and Canada, included several large developing countries such as Argentina, Brazil, Indonesia, and the Philippines) to purpose their interests. Furthermore, during and after the conclusion of the Uruguay Round, the formal accession of developing countries to the GATT and now the World Trade Organisation (WTO) has continued apace. Of the 134 members of the WTO in February 1999, some 70 per cent were developing countries. The United Nations classified 48 countries as least-developed (LLDCs). Within that group, 29 are members of the WTO, six are in the process of accession, and three are observers. Also, 18 countries have been identified as net-food-importing developing countries (NFIDCs).

Some Definitions

The LLDCs are identified by the United Nations General Assembly based on several criteria—income per capita, augmented physical quality of life index, and an index of economic diversification. As a group, they have a population of about 590 million people, with an income per capita about 4 per cent that of the world average (1996). Agricultural production per capita in LLDCs has been declining since the 1970s although the same indicator for all developing countries (mainly under the influence of China) has gone up by nearly 40 per cent in the same period. LLDCs represent a small fraction of world trade (less than

1 per cent for total and about 2 per cent for agricultural trade). They had a positive, although declining net agricultural trade balance until the mid 1980s, when it turned negative. Almost 20 per cent of their total imports are food items.

The 18 net-food-importing developing countries have been selected through a process within the WTO. They have a population of some 380 million people and an income per capita nearly five times that of the LLDC average, but still much lower than the world average. NFIDCs are a diverse group: four are upper-middle income countries; eight are lower-middle income; and six are lower income. Four of them had net food exports on average during 1995-97, but because they imported cereals they are included in the group. NFIDCs' per capita food production as share of both world and developing country averages has risen, although from very low levels.

Although the categories of 'developed' and developing' countries have important legal consequences under WTO rules, there are no formal definitions of either category. The process work through self-identification and negotiation with other member countries of the WTO.

Completing the Unfinished Agenda

In general, developing countries operate under what has been called 'special and differential treatment'. They face lower disciplines and enjoy longer time frames for implementing reforms. In the case of LLDCs, they are totally exempted from WTO commitments, and it has been agreed that developing and least-developed countries should receive special consideration for market access and technical and financial support. Also, during the Uruguay Round, concerns that liberalization of agricultural policies and trade could adversely affect the food imports of LLDCs and NFIDCs led participants to include several measures dealing with food security issues in the 'green box' of permitted domestic support—for instance, the formation of public stockholding and the provision of foodstuffs at subsidized prices. There was a ministerial decision in Marrakesh in April 1994 to deal with possible negative effects of agricultural trade reforms on the food security of LLDCs and NFIDCs. The decision was reemphasized at the 1996 ministerial meeting of the WTO in Singapore.

Export and Domestic Subsidies: While many developing countries have significantly reduced distorting domestic agricultural policies, the possible benefits that these countries and the world can enjoy are thwarted by the subsidies of developed countries. The Uruguay Round was a first step in imposing discipline on the unfair competition arising from subsidised agricultural exports, which hurts poor agricultural producers in developing countries irrespective of their net agricultural trade position. In the next negotiations, that first step should be completed with the elimination of export subsidies. Net-food-importing developing countries should also be interested in stricter disciplines on export taxes and controls that exacerbate price fluctuations in world markets.

Under the Uruguay Round agreement, there is still a lot of scope for the developed countries to use domestic subsidies, in addition to the use of export subsidies; to help their farmers. The developing countries should seek further disciplines in this regard, including, among other things, the elimination of exemptions under the 'blue box' (which allows farmers to receive some forms of direct payments that are considered to be trade distorting). Least-developed and developing countries, however, will still be allowed 'special and differential treatment' on these issues.

Market Access: If the developing countries are to succeed in diversifying their agricultural sectors, they need expanded access to markets in developed countries.

This includes increasing the volume of imports allowed under the current regime of tariff-rate quotas (TRQs, which replaced the previous system of rigid quotas with a combination of a quantitative quota and a high tariff for the eventual out-of-quota imports); making the administration of the TRQs more transparent and equitable; seeking further reductions in tariffs, particularly those still high in some key products; and completing the process of tariffication in the cases where exemptions were granted. Also, eliminating, or at least reducing, tariff escalation in non-agricultural products is important for developing countries: this practice undermines the possibilities of expanding production and exports of processed goods that use agricultural inputs, exploiting 'forward linkages' in the value-added chain.

What the Most Vulnerable Need

The special situation and concerns of least-developed countries and net-food-importing countries were recognized in a ministerial decision agreed upon at the completion of the Uruguay Round in 1993. These concerns include the preservation of adequate levels of food aid, the provision of technical assistance and financial support to develop the agricultural sector in those countries, and the continuation and expansion of financial facilities to help with structural adjustment and short-term difficulties in financing food imports. It is important to make food aid available in grant form, to target it to poor countries and social groups, and to deliver it in ways that do not displace domestic production in the countries receiving it. Badly managed food aid, or cheap food imports due to export subsidies, may just reinforce the bias of economic policies against the rural sector, with its negative impact on poor agricultural producers.

Volatility in agricultural prices must be monitored carefully. While expansion of world agricultural trade should limit overall fluctuations by spreading supply and demand shocks over larger areas, the decline in world public stocks as a percentage of consumption works in the opposite direction. Improving early warning of potential food shortages, lowering costs for food transportation and storage, and providing better targeted food aid programmes and financial facilities for emergencies are also issues that need to be addressed by countries participating in the coming round of negotiations.

The impact of changes in trade and agricultural policy on poorer consumers and producers in developing countries is a matter of debate. Some have argued that trade liberalization may hurt both groups. Others have answered that greater productivity and growth coming from better trade and sectoral policies should help generate employment and income, given a setting of adequate overall economic policies and properly functioning markets and social institutions.

Small producers will also be helped by the disciplines that the URAA is bringing to subsidised and dumped exports, while it allows the implementation of a variety of programmes aimed at poor producers or consumers, including stocks for food security purposes and domestic food aid for populations in need. The issue here is the adequate design and funding of domestic policies to achieve the intended objectives of agricultural growth and poverty alleviation, which most certainly will not be helped by trade-distorting interventions either in developed or developing countries.

In general, low-income developing countries and LLDCs should emphasize to the international community the importance of creating and expanding a supportive international trade and financial environment and of implementing an integrated framework for economic and social development, with functional and trade policies being an integral part of the strategy. Appropriate measures would include—in addition to the agricultural trade issues suggested here—the continuation and enhancement of the reduction of the external debt of Heavily Indebted Poor Countries (the HIPC initiative) and to further liberalization of trade in textiles.

But improved international conditions should go hand-in-hand with a better domestic framework in developing and least-developed, including stable macro-economic policies, open and effective markets, good governance, the rule of law, a vibrant civil society, and programmes and investments that expand opportunities for all, with special consideration for poor and disadvantaged groups.

Bringing Developing Countries into the Process

Developing countries, as small players in the global arena, should be interested and active participants in the design and implementation of international rules that limit the ability of larger countries to resort to unilateral action. Also, domestic legal and institutional frameworks in developing countries may be strengthened by the implementation of internationally negotiated rules that limit the scope for rent seeking and arbitrary projectionist measures. The developing countries as a group have much to gain from continued progress towards a transparent, rule-based, trading system in agriculture.

What are the requirements and skills for the developing countries to become effective members in the next WTO round? Any negotiation requires careful consideration of the legal, economic, and political dimensions that define the substance and possible evolution of the negotiations, as well as the diplomatic and negotiating techniques that may help in the attainment of the expected outcomes. Questions that need to be addressed include:

- What are the economic and social consequences of different WTO scenarios (quantitative estimation of impacts)? Knowing the impacts of alternative scenarios is crucial if developing countries are to represent their interests in the negotiation process.

- What are the legal issues being discussed (definition of obligations, exemptions, time frame, and so on)? Detailed knowledge of international trade law is crucial if developing countries are not to be 'shortchanged'. The devil is in the details.

- Looking at the political process, who are in the main actors and their interests and what type of alliances may drive the negotiations? Negotiators must understand the political economy of their own country and of other countries in the WTO if they are to negotiate effectively.

- With these elements, an adequate diplomatic and negotiating strategy must be defined and implemented.

Developing countries that have carefully considered all four components will be better prepared to participate effectively in the coming negotiations. Of course, limited financial and human resources act as an important constraint. However, developing countries may overcome some of the problems through collective action, for instance considering the creation of alliances with respect to their main export and import commodities and the markets they approach for their exports. An example is the Cairns Group. This approach could reduce the fixed costs of negotiations. Spreading them over groups of countries, allow a better use of scarce technical expertise, and improve the bargaining position of developing countries. It could also be in the interest of the OECD countries to deal with negotiating blocs, which represent a smaller number of negotiating positions, rather than with numerous separate countries. The negotiations would be much more efficient and balanced.

13

Strategies for Improved Water Management

Since all surface and sub-surface water related activities are closely interlinked through upstream-downstream relations in river-basins and aquifers, the whole river-basin must be considered in all water development policies and research.

River-basin authorities should be established, for both national and international basins, and should have regulatory powers over inter-sectoral allocation of water, enforcement of water quality standards, arbitration in disputes, and compensation procedures, International basin authorities should be composed partly of representatives of national basin authorities.

Water policies should encompass underground water, surface water, and direct utilization of rainfall, as an integrated continuum requiring common strategies.

Participation of Users and Stakeholders

Governments should strengthen stakeholder participation in the allocation and management of water resources by means of:

1. identifying and establishing water rights for all users, following economic, cultural, environmental and social considerations;

2. establishing compensation procedure (both national and international);

3. using impact assessment as a tool to identify livelihood rights and ecological considerations.

Stake-holder participation in water allocation and management should be used to reconcile multiple or contradictory objectives and to reduce conflicts over ownership or control of water.

Environmental Protection

To guide development actions and to monitor changes, basin-wide mapping of 'hot spots' (areas of special problems or vulnerability with respect to water quality) should be undertaken; sources of major pollutants should be identified and published. Water bodies which have special ecological or societal significance should be preserved.

Training of professionals for the water sector should develop new environmental dimensions, and should address such issues as; conservation of water; responses to water scarcity; disaster preparedness.

Research into the Impacts of Water Policy

The impacts of water policies, and policy changes, are not well understood. There is a need for structured research on impacts rather than reliance on anecdotal evidence.

Water development actions can have negative effects on equity, and can injure individuals or groups even while bringing wider benefits. There is a need for analysis of responses and solutions to these effects.

Externally defined systems of water rights will lack legitimacy and will fail to meet the needs of all users, if they do not take account of users' perceptions and customs. National research institutes should therefore develop and apply methodologies for assessing users perceptions and customary rights.

Capacity-Building for Integrated Management

The development of river-basin authorities will require the training of their personnel in: coordination; planing of systems for monitoring and evaluation; sensitisation to new responsibilities, integrated land and water resources users; use of modelling tools; implementation of guidelines; procedures for stake-holder participation.

Politicians and polity makers need to become more aware of issues concerning; water scarcity and water needs; equity of access to water; water rights; sustainability; service improvement; enhancing the performance of water systems; stakeholder participation.

External Assistance for Development

Partnerships or 'twinning' arrangements, should be promoted between new river basin authorities and existing, successful ones.

External donor assistance programmes should focus on users' groups, especially among the disadvantaged and the poor. They should take a basin perspective, recognizing existing institutional arrangements and inter-sectoral requirements. Within this frame work they should foster and encourage mechanisms for devolution of responsibilities, tasks, and authority to local level users' groups.

External help is needed to develop and evaluate water databases, agricultural knowledge systems, methodologies and processes for water development and evaluation, and other information areas.

Emergencies and Conflict Reduction

National governments and river-basin authorities should

develop contingency plans for the management of their responses to natural disasters, prolonged droughts, and other emergency situations in which normal procedures of water allocation and management may require temporary modifications.

Policies and actions in regard to shared water sources should be improved, with emphasis on consultation, coordination and sharing of information.

Irrigation Management—Facing the Challenge

Irrigated agriculture is up against an enormous challenge. India's population continues to grow at a tremendous rate. Over the next 10 years, there will be more extra mouths to feed. Water is becoming increasingly scarce for agriculture, with conflicting demands being made on limited supplies by the domestic and industrial sectors. The most attractive irrigation sites have already been exploited. Yet, irrigated agriculture will have to deliver average output increases of at least 3.5 per cent per year if future food demands are to be met in India.

Forty years ago in the 1950s and 1960s, India has worried about its capacity to produce sufficient food to feed its growing population. Episodes of food scarcity were not uncommon. It was dreaded that millions would die of hunger. Then dawned the miracle of the 'green revolution'. Cereal production increased by leaps and bounds, boosted by expanded irrigation, increased fertilizer application and modern crop varieties. The vigorous response in mobilizing financing towards boosting agricultural production paid off.

Given the importance of irrigation to the India's food supply and the vast resources already expanded on irrigation development, it is tragic that the actual performance of irrigation systems is so disappointingly low. In the post-green revolution era, it has become increasingly evident that the performance of irrigation systems, especially large-scale systems, is suboptimal whether measured in terms of achieving planned area targets or in terms of production potentials created by the physical works.

In many irrigation systems, the actual area irrigated is much less than the common area. Sharp inequalities in water supplies between farmers in the head reaches of irrigation systems and those located downstream is another manifestation of poor performance. Investigations in the Tungabhadra Irrigation Scheme reveal that the tail-end of a major distributory commanding 25 per cent of the total area, received approximately 20-40 per cent of the targeted discharge while the upper reaches got more than their share.

Lack of maintenance has caused many systems to fall into disrepair, further inhibiting performance. Over time, distribution canals have become silted up, increasing the likelihood of breaching, damage to outlets and lading to their build-up in the soil.

Successful Farmer-Managed Irrigation Systems

Farmers have long demonstrated their potential capability to manage irrigation systems efficiently. Farmer-managed irrigation systems, (FMIS) also known as traditional, indigenous, communal or peoples' systems, are often classified 'minor' or 'small-scale' irrigation systems, although they may be found to command areas of 15,000-20,000 hectares.

Many successful farmer-managed irrigated systems which have been functioning effectively for hundreds of years, represent a valuable, accumulated investment. They are also reservoirs of largely untapped irrigation management experience.

Research has revealed that FMIS contribute to the production of a significant portion of the subsistence food supply. Ground-water irrigation systems, often found in areas affected by drought, play a strategic role in promoting food security. In India ground-water development which is increasing in importance is predominantly farmer-managed. In this country as estimated 20 millions hectares are already under ground-water irrigation. When mismanaged though, ground-water irrigation could impact negatively on the environment.

In addition to the hectarage under farmer-managed ground-water irrigation, farmer-managed tank irrigation systems cover about 8.5 million hectares in this country. Farmer-managed irrigation systems have also allowed intensification of agriculture to partially meet the food needs of rapidly growing populations.

Little Awareness

Yet, despite the widespread interest in irrigation management turnover throughout India there is very little documentation about the processes used and the results obtained from irrigation management turnover. Many policy-makers do not know how to turn over management of their irrigation institutions in an effective way. They typically have very little, if any, awareness of the range of organisational options which may be suitable under different conditions. There is an urgent need, therefore, for a systematic, comparative assessment of the range of approaches being used, constraints to implementation and the impacts on performance of transferring management to non-governmental or farmers' organisations.

Successful irrigation in the future will be that which supports much higher levels of agricultural productivity, enhances responsiveness to more diversified and dynamic crop markets, stimulates more profitable irrigated agriculture for wide numbers of rural poor, substantially improves water use efficiency and supports the sustainable use of scarce land, biomass and water resources.

In the coming years the irrigation sector will be in ferment, with decision makers, agency managers and farmers needing better information and strategic processes to make intelligent choices in the management of irrigated agriculture.

15

Solving Conflicts Over Water Uses

There are few issues that have greater impact on the life of mankind and the planet as a whole than the management of our most important natural resource water. This has only been realised in detail more recently by the general public as well as by many planners and decision-makers. Around the world they have begun to appreciate the critical importance of a reliable water supply for their future survival and sustainable development. Rivers are lifelines in countries like India serving different uses such as transport, agriculture, fisheries, personal hygiene and others. Conflicts over use of water resources must be settled through better water management policies.

In many localities of the Earth water-related problems have become extremely acute, even critical. In some places they are the source of social instability and are a threat to international security. There is not the slightest doubt that, with further population increase and under the 'water business-as-usual' scenario, these problems will become ever major acute, thus creating ever more instability. After decades of water waste, water pollution, and inability to provide basic water services to the poor we must fundamentally change the way we think about and manage water. We have to realise that water can no longer be considered to be a cheap and plentiful resource, which can be used abused or squandered without much concern for further human welfare.

As fresh water is becoming scarce, that is when there is not enough water to satisfy all demands, competition develops. Besides the well known tensions at international level over limited water supplies there is an-increasing and in many cases a far more important competition over water arising within countries, between sectors. Such competition, for instance, occurs between farmers for irrigation water and between farmers and non-agricultural users of water such as cities (including industries and power plants) and environmental concerns (recreation, fish and other wildlife). Because of this increasing competition irrigated agriculture around the world, but especially in developing countries, faces important challenges in the coming decades, on the other hand, it has to provide a major share of the required increases in food and fiber production to meet the objectives of poverty alleviation and development. On the other hand. It is threatened by water shortages arising out of increasing competition from domestic, industrial and other sectors. This situation is further worsened by dwindling financial resources available for capital expenditures as the cost of new irrigation schemes increase.

Land under Irrigation

On a regional basis, it is estimated that for example around 60 per cent of the value of crop production in Asia is grown on irrigated land. The irrigated sector performs an essential task in meeting the basic food needs of billions of people. It has provided more than half of the two important basic staples and close to a third of all food crops. In the future, the irrigated sector will have to provide an even larger proportion of the total food output. The questions arises whether irrigated agriculture will be in the position to provide the extra food needed to feed a growing population despite an increasing water scarcity and inter sectoral competition. There is no general, no easy answer to this question. But the following implications are foreseeable.

The irrigation sector has to recognise that economic structures are dynamic and not static. The economics of many countries have undergone considerable change in the last decades. In connection with this change agriculture is losing its leading role in economic development and this is, besides other things, affecting the allocation of water resources between agriculture and other users. The same applies to many other countries around the world, especially developing countries in arid and semi-arid climates.

New Water Policies are Needed

Because of the growing competition for ever-scarcer water resources governments, and water authorities are forced to change water policies. Objectives of the new policies are demand-decreasing and demand-shifting. Irrigated agriculture, by far the dominant water user, will be strongly effected by such policy changes.

Water has an economic value in all its competing uses and should be recognised as an economic good. Agriculture will in future not any more get water free of charge as it did in the past. Farmers will have to pay for the water, and costs will be increasing steadily over the years to come:

There will be a reduction of the role of governments in rural water projects and an increasing importance of local user groups. Experience shows that an important solution to water related problems is to give users the responsibility for developing and managing the water resource. This requires that farmers become properly organised in water user associations able to discharge their new responsibilities, and what that they have security of tenure to the land they farm and irrigate.

During the recent decade growth in crop productivity in irrigated areas has slowed, and competition for water for non-agricultural uses has increased. These developments place strong demands to develop water resource policies to maintain growth in irrigated agricultural production:

- facilitate efficient allocation of water across sectors and final demands; and
- reverse the ongoing degradation of the water resource base, including the watershed irrigated land base, and water quality.

What water policies can lead to efficient increases in irrigated production while reducing resource degradation in the irrigated areas in developing countries and realising water for growing non-agricultural demands? What policies can be implemented to conserve water in non-agricultural uses to reduce competition between sectors?

Questions to be Answered

Among others the following question's still have to be answered:

1. What are the implications of growing competition between agricultural and non-agricultural uses of water for the availability and productivity of water in agriculture?
2. How to use regulations, water prices, pollution taxes and effluent charges to encourage water conservation and pollution control in industries and households?
3. What are the production, equity, employment generation and income impacts of alternative water allocation mechanisms in different agro-economic and scarcity environments? What investment and administrative costs are associated with different mechanisms? What is the impact of resource allocation methods on water use, cropping patterns, crop yields, and fertilizer and other input use, capital investments, farm income, and environmental degradation?
4. What is the connection between alternative allocative mechanisms and the environmental externalities caused by irrigation, including water logging, salinization, ground water recharge and ground water mining?

5. How will food security in low income countries be effected by changing water policies?

In the context of the growing competition for water it is important to enable policy makers to evolve a policy-mix which will result in efficient, equitable and environmentally sustainable allocation of water resources to user sectors. To achieve this, extensive research work is needed.

16

Fresh Water and the Environment

It is widely recognized that water is going to be one of the major issues confronting humanity at the turn of the century and beyond. We are facing a crisis as regards the quantity and quality of water supply, but we have yet to experience the full social and political impact of that crisis. The escalation in the population and the quest for continued development is leading to conflicting pressures on water resources. Such resources are the ultimate recipient of pollution from various socio-economic activities associated with urbanization, agriculture, mining and clearing of native vegetation. Pollution originating from human waste, especially where appropriate sanitation facilities are not available, or are located too close to water supply sources affects both surface water and ground water.

This makes water supply and health perhaps the most important issue for the larger proportion of the global population. Paradoxically, the demands for 'sustainable management' and increasing global population require more potable water from the declining available potable water base.

It is universally accepted that proper water administration is a critical component of sustainable development—that is, development that meets the needs of both present and future generations. Indeed, water is an essential factor in a large number of productive activities, of which one of the most important is the production of food by irrigation. This activity, accounts for two thirds of the water resources used by humanity. A supply of drinking water and sanitation in urban centers are crucial for preserving human health.

For some decades it has been known that the misuse of water resources is responsible for many important environment problems. For example, in many industrialized cities both surface water and ground water are seriously contaminated. This deterioration is a consequence of a range of human activities, sometimes in isolation, others over a large area or a long period of time. Among examples of the latter is modern agriculture, whether it uses irrigation or not, as a result of the intensive use made of mineral fertilizers and pesticides.

Water Shortage: Exaggeration, Reality or Bad Management?

Some of these problems have made news and have created the impression that water shortage will be one of humanity's big problems in the coming decades. Sometimes this feeling is due to genuinely manipulative publicity campaigns to justify the setting in motion of hydraulic mega projects which basically benefit a few large construction companies. The truth is that except for a handful of very specific cases, no problems of water shortage are to be found almost anywhere. On the other hand, cases of bad water management are not rare at all.

Basic Principles for Good Water Management

Good management of water resources—and of almost all other natural resources—must be based on the principles of solidarity, 'subsidiarity' and participation. The physical reality requires that these resources be considered a common heritage of humanity both now and in the future. By 'subsidiarity' we mean that water management should be as decentralized as possible: what one person or any minor social group can do should not be done by a higher authority. For example, what local government can do should not be done a regional, state or central government. Participation consists in water users playing as a large a part as possible in decisions affecting water, in keeping with each state's or country's social and cultural structure. Obviously this participation calls for a central cultural and technically knowledge—a hydrological education—on the part of the those users.

The need for participation by users is even greater in the exploitation of groundwater. In this case, users tend to extract water independently of one another. They often fail to realize, until there is a serious economic or environmental impact, that their pumping affects other people who rely on the same water supply as has happened.

Water shortage is rarely a serious problem: in fact, in some cases the problem is exaggerated to justify the construction of large works using taxpayer's money. On the other hand, the contamination of surface and groundwater tends to be a problem which rarely receives adequate treatment. Successful water management should be based on three basic principles: solidarity, subsidiarity and participation. The specific way in which these principles are applied will vary from one state or country to another, but the effectiveness of water management will depend in large measure on the hydrological education of the general public.

The universal way of obtaining fresh water is from rain. River systems are the results of the excess water that falls on dry land in the form of rain. On the one hand, rain water penetrates the permeable soils, saturates them and accumulates to form groundwater reservoirs, or aquifers, which can come to the surface in the form of springs. On the other hand, the water is absorbed by vegetation, which uses it for pumping minerals and then evaporates it by transpiration. Some rainwater is lost because it evaporates immediately on falling on impermeable surfaces like the asphalt of roads and cities. Running water courses finally flow over saturated soils, shaping the complex systems of the watersheds or river basins.

Since each basins natural system has developed gradually and has grown up according to the yearly distribution and fluctuations of rainfall, we have to appreciate that any large-scale project for redistributing water by means of pipes, as if it were gas or electricity, is a journey into the unknown. This is because it destroys the results of the work of shaping the climate, however transitory it might be.

Variable Volumes

All water supplies are of variable volume. Both the discharge of rivers and the level of lakes and aquifers depend on rainfall. As these resources are components of a larger system, the river basin, a reasonable policy would be to manage water resources according to the characteristics of each basin. This would require, first of all, a proper understanding of the system so as to adapt use and consumption to the existing supply. Conserving river systems as much as possible in their natural state is the best guarantee for the preservation of the landscape and of a constant supply. Groundwater reservoirs aren't canals, but are more like lakes, so that pollution leads to the build-up of a debt which is paid in years to come.

Consumption

Water consumption has increased in recent years as a result of not only population growth but also an increase in living standards. In the rural areas the introduction of new farming methods, the spread of irrigation and the excessive use of fertilizers and pesticides causes very high consumption—it is estimated that more than 2/3 of water consumption is used in irrigating. Agricultural pollution also endangers both surface water and aquifers, which receive water full of chemical products. Many cases of eutrophication, the enrichment of water by nutrients that accelerate the growth of algae, derive from the run-off of fertilizers. The practice of intensive stock-raising on farms with large numbers of animals also brings about these problems of over-consumption and pollution. Cleaning the stockyards requires large amounts of water which is then released into the environment with high concentrations of nitrogen.

As for industries they have in the past taken little care over water consumption and dumping, and in many areas the need for proper attention comes as something new. The best thing would be to make industry take its water at a point down-river from where it returns it or, better still, generalize the use of closed circuit systems based on the constant recycling and reusing of the same water.

As regards human consumption, the general attitude to cleanliness is based on diluting pollutants. One example is the success of the use of the Water Closet which involves diluting a few decilitres of urine in 10 or more litres of drinking water: quite a record in wastefulness.

Another aspect to be considered is the different quality of the water that falls on well formed soils from the water that falls on roads, cities, airports, suburbs and built-up areas and whose composition is less stable and 'worse' than that resulting from a more uniform interaction with mature soils. Remember that streets, roofs, communication routes, airports and built-up areas already cover a high proportion of the earth's land area and are still on the increase.

Purification techniques should be based especially on the natural processes that include biological activity. Otherwise-for example, if physico-chemical methods are used—there can be side-effects such as an excess of mud or sediments. The strategy to follow is to optimize operations in our use of water according to the discharge and to the distribution of contamination. A system in the form of a conduit or channel, such as a river, can respond relatively quickly. On the other hand, lakes and dams can only do so up to a point, because they show more inertia and irreversibility and take longer to clean.

Large lakes, not to mention the sea, might seem a good place to dump contaminating refuse, but they can't then be cleaned. This is the price we pass on the future generations: a comfortable attitude, but an unacceptable one.

17

Water: An Educational and Informative Approach

The most characteristic element of our planet is undoubtedly water. Indeed, more than two-thirds of the earth's surface is covered by water—the total volume representing almost 1,500 million cubic kilometers. About 94 per cent of this water is found in the oceans, almost 6 per cent is located underground and in glaciers whereas rivers, lakes, soil moisture and atmospheric vapour, which constitute the major source of drinking water, account for a mere 0.0221 per cent of the total volume.

Water is indispensable for all living organisms. Life, as we know it, is impossible without water. It is present in all aspects of our life—directly or indirectly next to the air we breathe and together with the soil that we live upon, water constitutes the most important part of our environment, our most precious resource. And yet, except in the arid or semi-arid regions of the world, its value is generally overlooked until some catastrophe—natural or man-induced—forces our attention to its worth. But even so, no sooner is the situation remedied than, more often that not, we revert to our old attitude.

The reason for this sort of indifference in undoubtedly attributable to the fact that, except in exceptional circumstances, water has always been considered as a 'gift of the God', as some thing that human beings are as naturally entitled to as the air they breathe. Its supply, however uneven, has always seemed inexhaustible because water has a natural regenerative cycle which, until the present century, was beyond human control or interference—of even proper comprehension. But the trend of social, political and economic evolution, notably in the past 200 years, with an increase of industry, agriculture, technology, and above all, a vertiginous population growth, as led to a dramatic revision of the age-old belief that no demands made by human populations on the natural resources of the planet are in the process of setting in motion vicious circles in the environment from which it is becoming increasingly difficult to extricate ourselves, not only as concerns the present, but far more important, for the future. Thus, the overuse-or abuse-of water resources has started affecting seriously not only the water cycle but the very nature of water in such a way that, in conjunction with other abuses of the environment, the results have been climate changes, droughts, flooding, desertification on the one hand and acid rain, water pollution and eutrophication on the other.

Actually, the problem of water is to be considered less in terms of quantity—though with a steeply increasing world population making increasingly heavier demands on a fixed quantity of water, one will sooner or later be confronted with this aspect of the problem too—than in terms of proper distribution of available resources taking into account sound management, stock-age and maintenance of quality. For among the major preoccupations of humanity in the coming years, adequate supply of freshwater to the teeming populations figures in the forefront. Between 1900 and 2000 water consumption will have globally increased ten-fold and though the share of agriculture, the major consumer of freshwater, is expected to drop significantly (from 90 per cent to 62 per cent approximately), that of industry and the cities will have increased enormously (approximately, from 6 per cent to 24 per cent and 3 per cent to 8 per cent respectively).

Given the current trend of societal evolution *i.e.* greater emphasis on industry and increasing migration towards the cities added to the global population boom, these figures are certainly cause for concern. Not only because of the damages caused to freshwater resources through the increasing use of fertilizers in the search to maximize agricultural production to cater to the increasing populations, but equally because the mushrooming of industries and urban concentrations are sources of increasing water pollution. Though the industrialized nations have more than their fair share of blame in this matter insofar as the current state of water pollution goes, for the future, it is in the developing world that lies the major source of concern. Lack of resources for adequate urban planning, the increasing role of industry in the search for economic solutions added to uncontrollable population pressures are already on the way to creating an explosive situation in a great number of developing nations with the available water supply becoming more and more inadequate in terms of quantity as well as quality. And when one considers the fact that around 80 per cent of all diseases are estimated to be water related, and that by the year 2000, 51 per cent of the world population will be urban based, one can hardly be accused of exaggeration in speaking of an explosive situation.

Attacking such a vast problem is no mean task. Water being at the very source of life, what concerns water concerns every aspect of life. Thus, be it climate change, pollution, desertification, deforestation, food production... or whatever other major environmental problem that humanity is confronted with today, water constitutes one of the prime factors. Managing our resources with care and intelligence for the use of present and future generations is a major responsibility which has to be shared by governments and the public alike, for no sector alone can deal efficiently with so vital a problem which affects not only the present but also the future of humanity. Again, as in the case of bio-diversity and climate change, the problem of water being a global problem, international cooperation is of utmost importance since activities in one part of the planet are likely to produce consequences in other regions of the world. Concerted action by the international community alone is capable of dealing effectively with a problem of such far-reaching consequences.

If our planet is to be saved from disaster—for in jeopardising our water resources we are guilty of nothing less than condemning life itself on our planet—we have to work for sustainable results: short-term plans for the present which will dovetail into medium-term ones for the coming generations without compromising the possibility, at the same time, of careful long-term planning to guarantee the future of the planet. In this, the part of environmental education and information of the people is fundamental. No strategy, no policy, no plan—be it ever so well prepared and implemented—can hope to succeed without the active and effective participation of the main actors—the people who must be properly educated and informed. For this age-old techniques, beliefs—mentalities must be brought in line with present day realities. People have to learn to think differently in order to veer from a course which, however right in the past, has been shown to be less than adequate for present conditions—and catastrophic for the future—and must therefore needs be altered.

Changing mentalities is neither an easy nor a rapid process. It is difficult to go back upon the accumulated experience of generations—even in the face of stark realities and scientific evidence. Moreover, when dealing with such global and fundamental issues as water, where even 'scientific evidence' tends to be stated in tentative terms, the task becomes more onerous. Add to this the fact that the problem presents itself most presently in developing countries which are equally subject to enormous economic pressures which tend to reduce the cope of possible solutions. We are thus faced with the enormous task of trying to change attitudes, values, mentalities of populations whose geographical, socio-cultural and economic conditions have already fashioned priorities other than those that would precisely permit them to overcome their difficulties in a sustainable manner. In other words, of persuading people to abandon traditional short-term strategies in favour of perhaps more unattractive but eventually sustainable, long-term practices.

A veritable Herculean labour—which can only be accomplished through information and education. And in particular, through environmental education and information whose avowed aim is precisely to develop the understanding, knowledge, skills and motivations leading to the acquisition of attitudes, values and mentalities which are necessary to deal effectively with environmental issues and problems. Sound and systematic environmental education of the people associated with concerted local, national and international action, is the only means to finding a sustainable solution to this problem. The ground has to be paved through adequate information on the subject followed by educative processes adapted to specific local conditions. For a uniform education, whether formal or non-formal, might perhaps do more harm than good as its rejection, due to its unsuitability in the light of local customs, beliefs, traditions... might only serve to reinforce the very attitudes that it seeks to change. In each region, each country, each locality the educative processes must correspond to the socio-cultural, historical, economic conditions of the people. Only then can we hope to arrive at the change in mentalities around the planet which, coupled with consistent, parallel support from national and international institutions, will lead to the safeguard of what is perhaps our most precious resource—Water.

18

Aquaculture: A Growing Danger to Environment in India

Aquaculture has made many Indian pond owners rich, but it is now the target of escalating protests by the region's farmers, fishermen and environmentalists alarmed by its adverse ecological and social effects. Opposition to aquaculture is fast gathering momentum in shrimp growing country like India.

Thousands of farmers have had their rice fields taken over by aquaculture companies and their rice crops destroyed by seepage of salt water from nearby ponds.

A strong grassroots movement in India has developed in the eastern coastal states, where local groups have organised themselves to prevent the building of shrimp ponds and to protect themselves from attacks perpetrated by aquaculture companies. In Andhra Pradesh, villagers have attacked aquaculture farms, uprooting the pumps and breaching the fences of the ponds. The activists recently won a Supreme Court order prohibiting new aquaculture projects in three states.

In India, fishing communities have seen sharp declines in their catch due to clearing of mangroves and river pollution caused by shrimp ponds set up along the coast. Fishermen have complained of loss of income. A 2.000-hectare ell farm destroyed valuable wetlands and caused many villages to suffer water shortage due to pumping of groundwater for the farm. Several hundred farmers have also protested against the state's acquisition of their rice fields for an aquaculture project. The farmers have taken their case to court.

The problem is not confined to India alone. The environmental group *Accion Ecologica* (Ecological Action) called for a consumer boycott of cultured shrimp because the aquaculture industry had destroyed most of the mangrove forests in world's coastal regions.

Opponents of shrimp farming are also strengthening their link-up with groups in other countries. At a meeting in Madras agreed to launch an international campaign against unsustainable aquaculture. Indian organisations also set up a National People's Alliance against the Shrimp Industry.

Although the ill effects of commercial aquaculture at the grassroots level is well known, there is still a misconception that it is environment-friendly and helps augment the food supply of poor communities. This could be because small farmers and fishermen in India have been practising aquaculture for centuries to improve their living conditions. But experts stress there is a huge difference between the traditional methods and those used by commercial firms.

Traditional aquaculture in India was usually small-scale, used low inputs and relied on natural tidal action for water exchange. In some countries like India, Bangladesh and Thailand, farmers adopted a rotating system, growing rice during part of the year and cultivating shrimp and fish the rest of the year. Chemicals, antibiotics and processed feeds were not used. In this low-yield, natural method, known as 'extensive aquaculture', the harvest was small but sustainable over long periods. The catch was only for family consumption or sold in local markets.

The modern method is larger in scale and intensive or semi-intensive in nature. Requiring more capital, it is owned and operated by commercial and often foreign owned companies. In intensive aquaculture, selected species are bred using a dense stocking rate. To maintain a large population and attain higher production efficiency, artificial feed, chemical additives and antibiotics are used. This, together with excrement from the shrimp, makes to wastewater from the ponds poisonous. Overcrowding also make the fish very prone to disease.

19

Water Problem in South India

Southern India's fast-growing urban areas and its farmers will collide over water allocation unless the government and water users take swift action. Urgently needed are measures that ensure efficient water use in cities and on farms, including regulation of groundwater withdrawal, restoration of traditional rain-collecting reservoirs, and experimentation with different cropping strategies.

In India's arid southern tip, typifies the plight of the country's extensive drylands. Scanty rainfall means residents often must get by on water drawn from small reservoirs and underground sources.

Extended dry periods are punctuated by intense monsoon bursts. When rains lash South India during the monsoons, only a fraction of the downpour is captured for later use. The rivers are seasonal and small compared to the Himalayan cataracts up north.

During the monsoon, these waters swell briefly to gigantic proportions recharging the water tables in their basins and then subside. In times of drought, when the rivers are narrow ribbons, water drawn from below ground sustains crops.

In the cities of South India the problems of water supply are exacerbated by antiquated or non-existent infrastructures that cannot keep up with frenetic growth.

Farmers in South India depend on irrigation to see them through the growing season. They draw their water from three sources: Government canals that bring water from rivers and dams, traditional reservoirs known as 'Tanks', and public and private wells often fitted with electric pumps. All three need to be made more efficient.

Besides being enormously expensive, large surface irrigation canals and dams are plagued by massive leaking and evaporation. Tanks collect rain water runoff behind small earthen dams. But these are falling into disrepair as farmers take advantage of low interest government loans and heavily subsidized electricity to switch to private wells fitted with electric pumps. Without maintenance, the tanks fill up with silt and hold less water.

As more wells are dug and as percolation tanks that once recharged underground water supplies fall into disuse, the water table is dropping at a rapid rate. Farmers who can afford to install powerful electric or diesel pumps on their wells are able to tap the retreating water table, but poorer farmers who raise their water by hand or cattle power from shallow wells often come up dry.

No regulations govern the amount of water a farmer can withdraw from a well, so underground reserves are available on a first-come, first-served basis. Add to this situation electricity rates based on the horsepower of pumps, not on the amount of electricity they consume, and there's no incentive to save water.

Any effort to limit groundwater extraction carries heavy political liabilities. Politicians are reluctant to risk raising the ire of farmers, who provide the majority of their votes.

Preventing overdraft where water tables are falling would be a first step toward correcting water troubles. While some areas face the prospect of rationing, others have an abundance of groundwater still to be tapped. Effective legislation must be based on detailed and accurate maps of groundwater supplies—something that doesn't currently exist.

They should also consider instituting higher user fees for its public canal projects and returning to its old policy of charging farmers according to the electricity they consume. Making farmers pay closer to the full price for water and electricity would provide revenue for digging communal wells equipped with electric pumps and overhauling the tank system, while providing an incentive for farmers to conserve.

Using tanks and wells together maximizes the effectiveness of both, since tanks take pressure off groundwater supplies and wells sustain crops during their final weeks of growth when tanks are low. Any efforts to limit the number of new wells dug has to include measures that provide something to those without water, otherwise only those who currently own wells will benefit from groundwater conservation.

Officials might also experiment with encouraging farmers to plant less water-demanding crops such as ragi, sorghum and pulses in place of cotton, sugarcane, bananas and spices.

If politicians act now, they still have a chance to craft measures that are fair than desperate. The time is nearing when water for drinking and irrigating crops will have to be culled from careful conservation rather than from the ground.

Population Growth and Fresh Water

Wherever population is growing, the supply of fresh water per person is declining. As a result of population growth, the amount of water available per person from the hydrological cycle will fall by 74 per cent between 1950 and 2050. Stated otherwise, there will be only one fourth as much fresh water per person in 2050 as there was in 1950. With water availability per person projected to decline dramatically in many countries already facing shortages, the full social effects of future water scarcity and difficult even to imagine. Indeed, spreading water scarcity may be the most under-rated resource issue in the world today.

Evidence of water stress can be seen as rivers are drained dry and as water tables fall. The Colorado River in the south-western United States now rarely reaches the sea. The Yellow River, the northernmost of China's two major rivers, has run dry for a part of each year since 1985, with the dry period becoming progressively longer. In 1997, it failed to make it to the sea for 226 days. The Nile, the largest river in the Middle East, the little water left when it reaches the sea.

Water tables are now falling on every continent, including in major food-producing regions. Among those where aquifers are being depleted are the U.S. southern Great Plains the North China

Plain, which produces nearly 40 per cent of China's grain; and most of India. Wherever water tables are falling today, there will be water supply cutbacks tomorrow, as aquifers are eventually depleted.

Some 70 per cent of the water pumped from under ground or diverted from rivers is used for irrigation. 20 per cent is used for industrial purposes, and 10 per cent is for residential use. Water use patterns vary widely by region. In Europe, for example, where agriculture is largely rainfed, water withdrawals are dominated by industrial use. In Asia, in contrast, irrigation accounts for 85 per cent of all water use.

As countries press against the limits of their water supplies, the competition among sectors intensifies. The economics of water use does not favour agriculture. One thousand tons of water can be used to produce one ton of wheat worth $200 or to expand industrial output by $14,000. This ratio of 70 to 1 explains why industry almost always wins in the competition with agriculture for water.

As the growing demand for water collides with the limits of supply, countries typically satisfy rising urban and residential demands by diverting water from irrigation. They then import grain to offest the loss of irrigation water. Since it takes at least 1,000 tons of water to produce a ton of grain, importing grain becomes the most efficient way to import water. North Africa and the Middle East—a region where population growth is rapid and every country faces water shortages—has become the world's fastest-growing grain import market during the 1990s. In 1997, the water required to produce the grain and other foodstuffs imported into the region was roughly equal to the annual flow of the Nile River.

In both China and India, the two countries that together dominate world irrigated agriculture, substantial cutbacks in irrigation water supplies lie ahead. The combination of the effects of aquifer depletion in key countries such as these and the growing diversion of irrigation water to non-farm uses in many countries makes it unlikely that these will be much, if any, increase in total irrigated area over the long term. Already the irrigated area per person has been slowly declining since 1978, falling from a historical high of 0.047 hectares per person to 0.045 hectares in 1996—a drop of 4 per cent. If the total irrigated area remains at roughly 263 million hectares until 2050, this key figure will fall to 0.028 hectares per person in 2050—declining by an additional 38 per cent. Such a shrinkage will pose a formidable challenge to the world's farmers.

About a billion people will be living in countries facing absolute water scarcity by 2025. These nations do not have enough water to maintain 1990 levels of food production per person from irrigated area, even with high irrigation efficiency, and to meet the needs for domestic, industrial, and environmental purposes as well. They will have to reduce water use in agriculture in order to satisfy residential and industrial water needs. The resulting decline in domestic food production will force them to import more food, assuming it is available. Although detailed water projections by sector for each country are not available for 2050, the number of water-deprived people will be far greater than in 2050 if the world continues on the U.N. medium population trajectory. The bottom line is that if we are facing a future of water scarcity, then we are also facing future of food scarcity.

21

South Asia Quarrels Over Water

Bangladesh, Bhutan, India and Nepal have at least three features in common; they are all situated on the southern slopes of the Himalayas, they have high levels of poverty and they are all rich in water resources. To these add a fourth: they frequently quarrel about water.

For decades, policy makers in the four countries have grappled with the problem of how to harness the rivers that flow from the mountains and serve the region's over-one billion people. Mostly, they have gone in for large dams and grand irrigation schemes. But though these promise a lot, their usefulness is being increasingly questioned by many independent engineers, scientists and social activists. And they are thought to contribute to water conflicts between countries.

Large hydro-power projects represent a development model that is biased toward industry and urban areas. And irrigation schemes favour rich farmers with large farmlands, most useful for water-and fertilizer-intensive cash crops that poor peasants cannot afford to grow. Because of their size, they also cause huge displacements-again, of poor farmers. Social activities say that while such projects may have increased overall food production, they have also increased the rich-poor gap in a region that is home to the world's largest number of absolute poor-people living on incomes of less than one dollar a day.

In tandem, many South Asian hydro-power engineers and economists say 'national' and top-down engineering solutions to water management ought to be replaced by a new 'basin-wide' or regional approach through environmental and socially benign projects.

One example of such 'national solutions' is the Indian Farrakka Barrage on the Ganges. Built near the Bangladesh border in order to divert water to the Calcutta port, it has led to problems with falling water tables and salinity downstream in Bangladesh. And although India and Bangladesh resolved the long-simmering dispute in 1996, Dhaka is now proposing another expensive Ganges Barrage to solve the problems created by the first barrage.

Of the four countries, Nepal has the highest per capita potential for hydropower generation—its narrow valleys and steep terrain mean water flows faster, which is good for power generation. But although Nepal could potentially turn its rivers into 'hydrodollars', Himalayan dams are expensive to build and the country's experience in dealing with India on joint river projects has been patchy. At the same time exporting power to India, where power cuts that can last up to 12 hours at a time in some cities, means foreign exchange for Nepal.

Because many Nepalese perceive past border irrigation projects as unfairly benefiting India, joint Indo-Nepal river projects are a political issue. A recent treaty to harness the Mahakali River on Nepal's western border with India is a case in point.

Although India and Nepal agree on equally sharing the 7,500 megawatts of power, from the project, negotiations have been stalled over Nepal's demand that it be compensated for downstream irrigation benefits to India. Nepal is not allowed to harness water in many of its rivers for its own use without India's concurrence these rivers flow into India, and Nepal is bound by past bilateral treaties.

As one official at Nepal's Water and Energy Commission put it: "What's in it for Nepal? Why should we write off benefits from the Mahakali for future generations of Nepalis? Things look even bleaker for larger joint projects like a planned 10,800 MW dam, on the Karnali River—the $ 10 billion Chisapani dam project is expected to displace some 60,000 Nepali farmers. In addition, India has already used up all the natural flow of the Karnali for its own irrigation downstream.

Things are different when it comes to Bhutan: India and the tiny Himalayan kingdom have sorted out their river projects without any major difficulty although experts say it is too early to access how much the Bhutanese will really benefit. According to an estimate by an Indian water consultant, Bhutan can generate up to 20,000 MW or power from its rivers.

Generating funds for large projects does not seem to be a problem. Traditionally, the World Bank has been one of the major financiers for dams worldwide having undertaken 400 projects involving dams since 1970. In 1988, India alone had about 473 dams under construction, with the Bank involved in 45 of them.

When the World Bank recently pulled out of some controversial mega projects, such as the Arun 111 dam in Nepal, multinational companies stepped in. In Nepal, three medium-sized hydropower projects worth more than 300 million dollars are being built with foreign direct investment.

FDIs are backed by organisations like Japanese-managed Global Infrastructure Fund (GIF) which is interested in putting money into such projects as the 270 metre-high Kosi High Dam in Nepal, the Ganges Barrage in Bangladesh and the mammoth 20,000 MW Dhihang dam in India. Funds also come from the Manila-based Asian Development Bank.

But many resource economists maintain that the chances of making colossal mistakes is higher with large projects. It is all a question of risk-management. Can our countries afford to take the risk of spending billions on doubtful projects that can go dangerously wrong?

A Rare and Precious Resource

Fresh water is a scarce commodity. Since it's impossible to increase supply, demand and waste must be reduced. But how?

Water is a bond between human beings and nature. It is ever-present in our daily lives and in our imaginations. Since the beginning of time, it has shaped extraordinary social institutions, and access to it has provoked many conflicts.

But most of the world's people, who have never gone short of water, take its availability for granted. Industrialists, farmers and ordinary consumers blithely go on wasting it. These days, though, supplies are diminishing while demand is soaring. Everyone knows that the time has come for attitudes to change.

Few people are aware of the true extent of fresh water scarcity. Many are fooled by the huge expenses of blue that feature on maps of the world. They do not know that 97.5 per cent of the planet's water is salty—and that most of the world's fresh water—the remaining 2.5 per cent—is unusable: 70 per cent of its is frozen in the icecaps of Antarctica and Greenland and almost all the rest exists to be form of soil humidity or in water tables which are too deep in the tapped. In all, barely one per cent of fresh water—0.007 per cent of all the water in the world, is easily accessible.

Over the past century, population growth and human activity have caused this precious resource to dwindle. Between 1900 and 1995, world demand for water increased more than six fold—compared with a three fold increase in world population seems to show that in overall terms there is enough water to go round. But in the most vulnerable regions, an estimated 460 million people (8 per cent of the world's population) are short of water, and another quarter of the planet's inhabitants are heading for the same fate. Experts say that if nothing is done, two-thirds of humanity will suffer from a moderate to severe lack of water by the year 2025.

Inequalities in the availability of water—sometimes even within a single country—are reflected in huge differences in consumption levels.

Scarcity is just one part of the problem. Water quality is also declining alarmingly. In some areas, contamination levels are so high that water can no longer be used even for industrial purposes. There are many reasons for this untreated sewage, chemical waste, fuel leakages, dumped garbage, contamination of soil by chemicals used by farmers. The worldwide extent of such pollution is hard to assess because data are lacking for several countries. But some figures give an idea of the problem. It is thought for example that 90 per cent of waste water in developing countries is released without any kind of treatment.

Things are especially bad in cities, where water demand is exploding. For the first time in human history, there will soon be more people living in cities than in the countryside and so water consumption will continue to increase. Soaring urbanization will sharpen the rivalry between the different kinds of water users.

Curbing the Explosion in Demand

Today, farming uses 69 per cent of the water consumed in the world, industry 23 per cent and households 8 per cent. In developing countries, agriculture uses as much as 80 per cent. The needs of city-dwellers, industry and tourists are expected to increase rapidly, at least as much as the need to produce more farm products to feed the planet. The problem of increasing water supply has long been seen as a technical one, calling for technical solutions such, as building more dams and desalination plants. Wild ideas towing chunks of icebergs from the poles have even been mooted.

But today, technical solutions are reaching their limits, economic and socio-ecological arguments are levelled against building new dams, for example: dams are costing more and more because the best sites have already been used, and they take millions of people out of their environment and upset eco-systems. As a result, twice as many dams were built on average between 1951 and 1977 than during the past decade.

Hydrologists and engineers have less and less room for manoeuvre, but a new consensus with new actors is taking shape. Since supply can no longer be expanded—or only at prohibitive cost for many countries—the explosion in demand must be curbed along with wasteful practices. As estimated 60 per cent of the water used in irrigation is lost through inefficient systems, for example.

Economists have plunged into the debate on water and made quite a few waves. To obtain 'rational use' of water, *i.e.* avoiding waste and maintaining quality, they say consumers must be made to pay for it. Out of the question, reply those in favour of free water, which some cultures regard as 'a gift from heaven'. And what about the poor, ask the champions of human rights and the right to water? Other important and prickly questions being asked by decision-makers are how to calculate the 'real price' of water and who should organise its sale.

The State as Mediator

The principle of free water is being challenged. For many people, water has become a commodity to be bought and sold. But management of this shared resource cannot be left exclusively to market forces. Many elements of civil society—NGO's, researchers, community groups—are campaigning for the cultural and social aspects of water management to be taken into account.

Even the World Bank, the main advocate of water privatization, is cautious on this point. It recognizes the value of the partnerships between the public and private sectors which have sprung up in recent years. Only the state seems to be in a position to ensure that practices are fair and to mediate between the parties involved—consumer groups, private firms and public bodies. At any rate, water regulation and management systems need to be based on other than purely financial criteria. If they aren't, hundreds of millions of people will have no access to it.

23

Solutions for a Water-Short World

As populations grow and water use per person rises, demand for fresh water is soaring. Yet the supply of fresh water is finite and threatened by pollution. To avoid a crisis, many countries must conserve water, pollute less, manage supply and demand, and slow population growth.

Caught between growing demand for fresh water on one hand and limited and increasingly polluted water supplies on the other, many developing countries face difficult choices. Populations continue to grow rapidly. Yet there is no more water on earth now than there was 2,000 years ago, when the population was less than 3 per cent of its current size. Rising demands for water for irrigated agriculture, domestic (municipal) consumption, and industry are forcing stiff competition over the allocation of scarce water resources among both areas and types of use.

Today 31 countries, accounting for under 8 per cent of the world population, face chronic freshwater shortages. By the year 2025, however, 48 countries are expected to face shortages, affecting more than 2.8 billion people—35 per cent of the world's projected population. Among countries likely to run short of water in the next 25 years are Ethiopia, India, Kenya, Nigeria, and Peru. Parts of other large countries, such as China, already face chronic water problems.

In much of the world polluted water, improper waste disposal, and poor water management cause serious public health problems. Such water-related diseases as malaria, cholera, typhoid, and schistosomiasis harm or kill millions of people every year. Overuse and pollution of water supplies also are taking a heavy toll on the natural environment and pose increasing risks for many species of life.

What Can be Done?

It may already be too late for some water-short countries with rapid population growth to avoid a crisis. Many other countries can avoid the coming crisis if appropriate policies and strategies are formulated and acted on soon. Whether water is used for agriculture, industry, or municipalities, there is much room for conservation and better management. Effective strategies must consider not only managing the water supply better but also managing demand better.

To avoid catastrophe over the long term, it is also important to act now to slow the growth in demand for fresh water by slowing population growth. Currently, in many developing countries millions of people want to plan their families and to use contraception. Family Planning programmes have played an important role in assuring individual reproductive health and in reducing national fertility levels. Continuing and expanding these programmes also can help assure that population growth eventually slows to sustainable levels in relation to the supply of freshwater.

Toward a Blue Revolution

The world needs a Blue Revolution to conserve and manage freshwater supplies in the face of growing demand from population growth, irrigated agriculture, industries, and cities—

just as the Green Revolution transformed agriculture in the 1960s. A Blustet Revolution will require coordinated responses to problems at local, national, and international levels.

Locally led initiatives show that water can be used much more efficiently. When communities manage freshwater resources efficiently, they also manage other natural resources better, improve sanitation, and reduce disease. At the national level, especially in water-short regions with dense populations, adopting a watershed or river-basin management perspective is a needed alternative to uncoordinated water-management policies by separate jurisdictions. At the international level countries that share river basins can fashion workable policies to manage water resources more equitably. Development agencies need to focus more on assuring the supply and management of freshwater resources and on providing sanitation as part of development and public health programmes.

A water-short world is an inherently unstable world. As the next century dawns, water crises in more and more countries will present obstacles to better living standards and better health and even bring risks or outright conflict over access to scarce freshwater supplies. Finding solutions should become a high priority now.

24

The Coming Water Crisis

Freshwater is emerging as one of the most critical natural resource issues facing humanity. The world's population is expanding rapidly. Yet there is no more freshwater on earth now than there was 2,000 years ago, when the population was less than 3 per cent of its current size.

Water is, literally, the source of life on earth. The human body is 70 per cent water. People begin to feel thirst after a loss of only 1 per cent of bodily fluids and risk death if fluid loss nears 10 per cent. Human beings can survive for only a few days without freshwater. Yet, in a growing number of places people are withdrawing water from rivers, lakes, and underground sources faster than they can be recharged—"unsustainably mining what was once a renewable resource," as one researcher puts it. Currently, 31 countries—mostly in Africa and the Near East—face water stress or water scarcity.

Population growth alone will push an estimated 17 more countries, with a projected population of 2.1 billion, into these water-short categories within the next 30 years. By the year 2025, 48 countries, with more than 2.8 billion people—35 per cent of the projected global population in 2025—will be affected by water stress or scarcity. Another nine countries, including China and Pakistan, will be approaching water stress.

Beyond the impact of population growth itself, the demand for freshwater has been rising in response to industrial development, increased reliance on irrigated agriculture, massive urbanization, and rising living standards. In this century, while world population has tripled, water withdrawals have increased by over six times. Since 1940 annual global water withdrawals have increased by an average of 2.5 per cent to 3 per cent a year compared with annual population growth of 1.5 per cent to 2 per cent. In developing countries over the past decade water withdrawals have been increasing by 4 per cent to 8 per cent a year.

Moreover, the supply of freshwater available to humanity is shrinking, in effect, because many freshwater resources have become increasingly polluted. In some countries lakes and rivers have become receptacles for a vile assortment of wastes, including untreated or partially treated municipal sewage, toxic industrial effluents, and harmful chemicals leached into surface and ground waters from agricultural activities.

Caught between finite and increasingly polluted water supplies on one hand and rapidly rising demand from population growth and development on the other, many developing countries face uneasy choices. The lack of freshwater is likely to be one of the major factors limiting economic development in the decades to come.

Slowing Demand, Conserving Supplies

To avoid a water crisis, particularly in water-short countries with rapid population growth, it is vital to slow the growth in demand for water by managing the resource better, while at the same time slowing population growth as soon as possible. Family Planning programmes play an important role not only for individual reproductive health but also for sustainability of the use of freshwater and other natural resources in relation to population size.

As population grows, so does demand for freshwater for food production, household (municipal) consumption, and industrial uses. The availability of freshwater limits the number of people that an area can support and affects standards of living. In turn, population growth and density typically affect the availability and quality of water resources in an area, as people attempt to assure their water supply by digging wells, constructing reservoirs and dams, and diverting the flow of rivers. If needs consistently outpace available supplies at some point overuse of water leads to the depletion of surface and groundwater resources, triggering chronic water shortages.

Scarce and unclean water supplies are critical public health problems in much of the world. Polluted water, water shortages, and unsanitary living conditions kill over 12 million people a year.

Competition for freshwater supplies breeds social and political tensions. River basins and other water bodies do not respect national borders. For example, one country's use of upstream water often subtracts from the supply available for use of downstream countries. As the 21st century dawns, there is a growing risk that wars will be fought over access to freshwater supplies.

If a crisis is to be averted, the world's overuse and misuse of freshwater must end as soon as possible. We cannot afford to keep wasting and fouling our precious supplies of freshwater. Increasingly, human activities are altering the flow of water and drawing down freshwater supplied faster than they can be replenished. Throughout the world enormous amounts of water are wasted due to inappropriate agricultural subsidies, inefficient irrigation systems, leaky municipal pipes, improper pricing of municipal water, poor watershed management, and other imprudent practices. It is time for widespread conservation measures, effective water management policies, and growing attention to assuring freshwater supplies and decent sanitation as part of development and public health projects.

25 Water: Will be There Enough?

Humankind has a special relationship with water. In every civilization, the most ancient traditions associate this precious resource with the origins of life, purification and regeneration. Far from being a mere raw material such as oil, water is vital for life, indispensable to the economy and so rich in symbolic value that it triggers passionate responses. All the computers in the world will never be able to express the real perception of the value of water or codify the interactions between it and peoples.

For decades, experts have been making grim forecasts that the Earth will start running out of water and that conflicts over this precious liquid will erupt into wars. The situation is indeed alarming.

How much water is there in the earth's reserves? Highly expensive probes have been sent to the Moon, Mars and the satellites of Jupiter and Saturn to findout whether there is water on them, but we will still lack accurate data about the earth's hydrological resources. Such information would help to provide a cleaner picture of the future and, especially to foresee the global repercussions of demographic and climate change.

One thing we know about water is that there is plenty of it. The total volume is put 1.4 billion cubic kilometers—which could be imagined as a 2,650-metre-deep layer of liquid evenly disturbed over the entire surface of the planet. But 98 per cent of it is salt water, mainly in the oceans and seas. Most of the earth's fresh water is trapped in the polar ice caps, less than 1 per cent of it is available in lakes, rivers and shallow, easily-accessible acquifers. These water resources are constantly in flux. Water from the oceans and land evaporates into the atmosphere before falling again as rain or snow, nourishing plants and swelling rivers that flow into the sea. *It also seeps* through the ground and percolates down to acquifers. Very deep groundwater, known as fossil water, is impervious to seepage and not renewable.

In the industrialized countries, all you have to do is turn a tap and before you know where you are you've used a considerable amount of water up to 600 litres per person a day in the United States. In not developing countries, where shanty-towns on the edges of cities are crowded with growing numbers of migrants from the rural areas, a spigot and two litres of water a day are a luxury.

Over 1.3 billion people received improved drinking water services and some 750 million got better sanitation facilities during the International Drinking Water Supply and Sanitation Decade (1980-1990). Approximately 1.2 billion people still have no access to drinking water and 2.9 billion lack sanitation. The resulting water-borne diseases take the lives of five million people a year, most of them children.

Farming and manufacturing account for most of the world's water consumption, far outdistancing human needs of the 3,240 km^3 of fresh water drawn every year, only 8 per cent are used for human consumption. Each year fewer than ten countries use 60 per cent of the world's 40,000 billion m^3 of surface and ground water. Lastly, per capita consumption rises with the standard of living, ranging from 260 litres a day person in Israel to 200 in Europe 70 for a Palestinian on the West Bank and 30 in Africa.

The Dangers of Irrigation

Demand runs highest in places where irrigation is indispensable, such as central Asia, Iraq, Iran, Pakistan, Madagascar and also in some industrially developed countries such as the United States. Farming accounts for two-thirds of the total water resources used by humans—a figure that rises to 80 per cent in the Southern countries. Developing countries consume twice as much water per hectare to irrigated land as industrialized nations, yet their production is three times lower.

Because of the heat, half the water, evaporates in storage areas or when flowing through open-air irrigation canals, Poorly conceived irrigation rejects lead to deterioration of the soil.

The first is Pakistan, during the first half of the twentieth century, 10 million hectares were abundantly irrigated in the Indus plain. Waterlogging caused by irrigation, combined with a high rate of evaporation, has led to salinization of the soil, making it unproductive. The second example, the Aral Sea in the former Soviet Union, is different but the result is the same. Much of the water from the Syrdarya and Amu Darya rivers that flow into the huge lake has been diverted to feed 1,80,000 kms of irrigation canals, only 12 per cent of which have been made watertight. The rivers' flow is considerably restricted and the Aral Sea is drying up. Irrigation for agricultural purposes is expensive. To make it profitable farmers must use massive amounts of pesticides, herbicides and fertilizers on increasingly exhausted soil. The impact of pesticides on health has been over looked. The child morbidity and mortality rates are among the world's highest.

26

End of Controversy on Large Dams?

Was it worth the effort? The energy, the time, the money invested? Quite a number of people probably put this question to themselves on the 16th of November 2000 when Nelson Mandela launched the final report of the World Commission on Dams (WCD) in London. In an extraordinary process that lasted two and a half years, dam proponents and opponents worked intensely together. Twelve commissioners had tried hard to come up with a consensus on the effects of large dams in the past, and with recommendations for future sustainable planning on water and energy issues. And surprisingly enough: the Commission succeeded. The cost of ten million dollars was financed by governments, international agencies, the private sector, NGOs and various foundations—this is also a novelty. Dam-affected people, non-governmental organisations, companies, consultants, politicians etc. contributed to the processes with their know-how and by giving support and additional resources to the Commission's work.

The establishment of the multi-stakeholder commission was the result of a growing and aggravating controversy about the social, ecological and—often enough also economic costs of large dams.

Global Review Shows Faulty Planning Processes

The WCD Global Review of large dams proves that there were good reasons for massive resistance against large dams in the past. As the report says: "In too many cases an unacceptable price has been paid... especially in social environmental terms, by people displaced, by communities downstream, by taxpayers and by the natural environment" while funders of large dams, political institutions, consultants and companies claimed in recent years that they had learned from their faults and improved their performance, the WCD highlights that, while policies and assessment procedures have been improved, it appears that business-as-usual too often continued to prevail:

- Even in the 1990s, impacts on downstream livelihoods were not adequately assessed or accounted for in the planning and design of large dams.
- Participation and transparency in planning processes for large dams was neither inclusive nor open, and while actual change in practice remains slow, even in the 1990s, there is increasing recognition of the importance of inclusive processes.
- Where opportunities for the participation of affected people and the undertaking of environmental and social impact assessment have been provided they often occur late in the process, are limited in scope and even in the 1990s their influence in project selection remains marginal.

Dams Hindered Human Development

The WCD states that "dams have made an important and significant contribution to human development, and the benefits derived from them have been considerable". A viewpoint that dam-affected people and NGOs hardly share, having in mind the experiences of the past. Medha Patkar, WCD Commissioner and activist in the struggle to save the Narmada river in India wrote in her comment to the Report: "Within the value framework the Commission propagates—equity, sustainability, transparency, accountability, participatory decision-making and efficiency—

large dams have not helped attain, but rather hindered 'human development'.

The WCD also puts an end to the old viewpoint that the violation of human rights and the social costs that some have to pay can be justified by the benefits for the others. The idea that society as a whole would profit from so-called development that were benefiting from 'trickle down effect'. In fact this development model tends to aggravate social inequalities and encourage environmental destruction leaving the rich better off, but the poor more marginalised and resentful. It is not a sustainable model.

It's Not Just About Dams

Although it is called the World Commission on Dams the issues before the Commission are much broader. The discussion about larger dams is inevitably linked with the need to find solutions for supplying water and energy in the future. To do this in a sustainable way is one of the challenges of our time.

The WCD defined 5 core values that are based on internationally accepted norms like the Universal of Human Right to Development and the Rio Declaration on Environment and Development:

- Equity
- Efficiency
- Participatory Decision-making and Accountability

These values are not really new and some of them have been agreed upon decades ago, but the Global Review of the WCD showed that 'in real life' we are still far from implementing them.

The rights and risks approach of the WCD must be mentioned as an important tool: while funders and dam-builders talk a lot about their (mostly) financial risks, risks of the dam-affected people were not a big issue in the past. The Commission makes an important distinction: while the former take a voluntary risk and have the possibility to decide on whether or not they want to take it, the latter take involuntary risks and so far had hardly any option in deciding on whether or not they are willing to incur them. Rights must not be violated in order to serve the needs of other people. Respecting human rights is the minimum basis for discussion and is not negotiable.

To get better results in the future the WCD defined seven strategic priorities for decision-making:

- Gaining Public Acceptance
- Comprehensive Options Assessment
- Addressing Existing Dams
- Sustaining Rivers and Livelihoods
- Recognising Entitlements and Sharing Benefits
- Ensuring Compliance
- Sharing Rivers for Peace, Development and Security

According to the WCD there are several fundamental strategic points in the decision-making process. On point is right at the beginning of the planning stage: if the discussion about sustainable energy and resources management is carried out in the transparent, open and participatory process, further steps are much more likely to be accepted by all parties and will help prevent conflict at later project stages. According to the WCD there are several fundamental strategic points in the decision making process. One point is right at the beginning of the planning stage: if the discussion about sustainable energy and resources management is carried out in a transparent, open and participatory process, further steps are much more likely to be accepted by all parties and will help prevent conflicts at later project stages.

The strategic priority 'Addressing existing dams' includes (among other policy principles) the identification and assessment of outstanding social issues associated with existing large dams. Considering that between 40 to 80 million people have been displaced by large dams (a lot more have been directly or indirectly affected) this is a difficult task. Financial institutions, development agencies and companies are a lot more interested in looking at the future than dealing with the complex problems of the past. But there is no way out. Dam-affected people of the past along with future dam-affected people are unlikely to accept new dams and believe in what is being promised while the legacy of the past is being forgotten. To solve these problems is a condition for further constructive discussions if not a mere moral obligation. How can those responsible for planning future dams think of gaining public acceptance if not by showing that they mean what they say?

The recommendations of the WCD are of fundamental importance, because they were agreed upon in a multi-stakeholder process. In other words, they were made by representative of industry as well as those from dam-affected people, by members coming from industrialized countries as well as those from developing countries. Within the Commission, it was possible to integrate the most diverse views and to come to a consensus.

No more Dams?

Although in their assessment the evidence would have allowed even stronger recommendations, NGOs and people's movements welcomed the report and asked for immediate implementation. Other stakeholders do not seem to be very enthusiastic about it. Quite a few representative from industry and investors seem to fear that the implementation of the WCD recommendations would lead to an abrupt stop in dam building.

All appreciate that the WCD report vindicates many concerns raised by NGO campaigns. Given the role of financial institutions in funding large dams and in the WCD process, and based on the WCD report's recommendations, all has to call on all public financial institutions, including the World Bank, the regional development banks, the export credit agencies and bilateral aid agencies, to take the following actions:

- All public financial institutions should immediately and comprehensively adopt the recommendations of the World Commission on Dams, and should integrate them into their relevant policies, in particular those on water and energy development, environmental impact assessment, resettlement, and public participation. In particular, as recommended by the WCD, no project should proceed without the free, prior and informed consent of indigenous people, and without the demonstrable acceptance of all those who would be affected by the project.

- All public financial institutions should immediately establish independent transparent and participatory reviews of all their planned and ongoing dam projects. While such reviews are taking place, project preparation and construction should be halted. Such reviews should establish whether the respective dams comply, as a minimum, with the recommendations of the WCD. If they do not, project should be modified accordingly or stopped altogether.

- All institutions which share in the responsibility for the unresolved negative impacts of dams should immediately initiate a process to establish and fund mechanisms to provide reparations to affected communities that have suffered social, cultural and economic harm as a result of dam projects.

- All public financial institutions should place a moratorium on funding the planning or construction of new dams until they can demonstrate that they have complied with the above measures.

Water Facts and Findings on Large Dams?

Today, around 3800 km^3 of fresh water is withdrawn annually from the world's lakes, rivers and aquifers. This is twice the volume extracted 50 years ago. World population has passed 6 billion. Projections say that it will reach a peak of between 7.3 billion and 10.7 billion around 2050 before total population begins to stabilise or fall.

50 litres per person per day (or just over $18.25m^3$ a year) covers basic human water requirements for drinking, sanitation, bathing and food preparation. In 1990, over a billion people had access to less than 50 litres of water a day. Agriculture accounts for about 67 per cent of withdrawals, industry uses 19 per cent and municipal and domestic uses account for 9 per cent.

One third of the countries in water stressed regions of the world are expected to face severe water shortages this century. By 2025 there will be approximately 6.5 times as many people—a total of 3.5 billion living in water-stressed countries. By the end of the 20th century, there were over 45,000 large dams in over 150 countries. The average large dam today is about 35 years old. Since average construction periods generally range from 5 to 10 years, this indicates a worldwide annual average of some 160 to 320 new large dams per year.

During the 1990s, an estimated $ 32-46 billion was spent annually on large dams, four-fifths of it is developing countries. Of the $ 22-31 billion invested in dams each year in developing countries, about four-fifths was financed directly by the public sector. About one-fifth of the world's agricultural land is irrigated, and irrigated agriculture accounts for about 40 per cent of the world's agricultural production.

Half the world's large dams were built exclusively or primarily for irrigation, and an estimated 30 to 40 per cent of the 271 million hectares of irrigated lands worldwide rely on dams. Dams are estimated to contribute to 12-16 per cent of world food production. Hydropower currently provides 19 per cent of the world's total electricity supply, and is used in over 150 countries with 24 of these countries depending on it for 90 per cent of their supply.

Floods affected the lives on average, of 65 million people between 1972 and 1996, more than any other type of disaster, including war, drought and famine. There are 261 water sheets that cross the political boundaries two are more countries. A number of key international rivers lack a basin-wide agreement that defines a process for establishing equitable use between riparian States.

Cost Effectiveness

Cost performance data confirms that large dam projects often incur substantial capital cost overruns. The average overrun was half again as much as the projected cost. The bulk of hydropower projects have delivered power within a close range of pre-project targets but with an overall tendency to fall short of targets.

At current rates, water fees are rarely sufficient to recover both capital and recurrent costs for water supply systems in many developing countries. Growing concern over the cost and effectiveness of large dams are related structural measures as long-term responses to floods has led to support for integrated flood management as opposed to flood control.

Multi-purpose schemes are inherently more complex, and many experience operational conflicts that contribute to under-performance on financial and economic targets. Substantive evaluations of project performance are few in number, narrow in scope, and poorly integrated across impact categories and scales with development becomes a process governed by negotiated agreements is critical to positive resettlement and rehabilitation.

Poor accounting in economic terms for the social and environmental costs and benefits of large dams implies that the true economic efficiency and profitability of these schemes remains largely unknown.

The direct adverse impacts of dams have fallen disproportionately on rural dwellers, subsistence farmers, indigenous people, ethnic minorities, and women. Where costs and benefits accrue to different groups, the standard procedures for adding up and discounting the expected costs and benefits do not provide an appropriate measure of changes in social welfare.

Financing

Almost 2 billion people, both urban and rural poor, have no access to electricity at all. Most efficiency measures and technologies are cost-effective at today's electricity prices and the use of full environmental and social costing of electricity supply options makes them even more so. Among advanced technologies in research and development, micro turbines and fuel cell show the greatest near and mid-term promise.

Total financing for large dams from multi-lateral and bi-lateral development banks comes to more than $4 billion annually at the peak of lending during 1975-84. Although the proportion of investment in dams directly financed by bi-laterals and multilateral was perhaps less than 15 per cent. The total investment in dams by the multilaterals and bilaterals since 1950 is approximately $ 125 billion.

Ecological Costs

Dams, inter basin transfers, and water withdrawals for irrigation have fragmented 60 per cent of the world's rivers. As a physical barrier the dam disrupts the movement in upstream and downstream species composition and even species loss. In Africa, the changed hydrological regime of rivers has adversely affected floodplain agriculture, fisheries, pasture and forests that constituted the organising element of community livelihood and culture.

Problems may be magnified as more large dams are added to a river systems, resulting in an increased and cumulative loss of natural resources, habitat quality, environmental sustainability and ecosystem integrity. Good site selection, such as not building large dams on the main-stem of a river system, and better dam design also played significant roles in avoiding or minimising impacts. The economic appraisal techniques such as risk and distributional analysis were still mandated for only 20 per cent of large dam projects even in the 1990s.

Social Costs

At the planning and design stage, an important social impact is the delay between the decision to build a dam and the onset of construction. This can result in communities living for decades starved of development and welfare investments.

The overall global level of physical displacement could range from 40 to 80 million. In India and China together, large dams could have displaced between 26-58 million people between 1950 and 1990. Little or no meaningful participation of affected people in the planning and implementation of dam projects—including resettlement and rehabilitation has taken place.

Empowering people, particularly the economically and socially marginalised by respecting their rights and ensuring that resettlement.

A Breakthrough in the Evolution of Large Dams?

Back to the Negotiating Table

"The problem is not the dams. It is the hunger. It is the thirst. It is the darkness in a township". With these plain words, former South African President Nelson Mandela summed up the World Commission on Dams (WCD's) motives in a speech at the presentation praising its work. His position is similar to that of the many other representatives of the South who were present: he demands a right to development. But Ms Medha Patkar, of India, a WCD Commission member and founder of the Struggle to Save the Narmada River (Narmada Bachao Andolan) anti-dam movement, takes a contrary view. "The problems of the dams are only a symptom of the larger failure of the unjust and destructive dominant development model," she says. We need to challenge "the forces that lead to the marginalisation of a majority through the imposition of unjust technologies like large dams".

Stragnation in Dam Building

The WCD is a unique experiment in reaching consensus. It

began in April 1997 when with the support of the World Bank and the World Conversation Union (IUCN), 39 representatives of diverse interests met at a workshop in Gland, Switzerland. At this point the various positions of the participants from governments, the private sector, international financial institutions, civil society organisations and affected people were cast in stone. The Manibeli Declaration in June 1994 of 326 activist groups from 44 countries had called for an immediate moratorium on World Bank funded large dams until a comprehensive, independent review of all Bank funded projects had been conducted. International financial institutions were actually no longer able to fund further large dams in the face of public criticism. Enervated by steadily growing protests despite continual tightening of social and environmental standards, the institutions' representatives wearily likened the dispute to a football match in which somebody kept moving the goalposts. But one proposal to emerge from the meeting in Gland was for all parties to work together in establishing the World Commission on Dams.

The WCD began its work in May 1998 under the chairmanship of Prof. Kedar Asmal, then South Africa's Minister of Water Affairs and Forestry. Its 12 members were chosen to reflect regional diversity, expertise and stakeholder perspectives. But the Commission ran the risk of failure right from the start due to their confrontational attitudes. The members, who spent 2 years 6 months jointly organising hearings, consultations and case studies and analysing more than 100 existing large dams, could not be more disparate.

The Current Situation

A large dam is a dam with the height of 15m or more from the foundation. If dams are 5-15 metres high and have a reservoir volume of more than three million cubic metres, they are also classified as large dams. Using this definition, there are more than 45,000 large dams around the world, almost half of them in China. They were built in the 20th century to meet the constantly growing demand for water and electricity. On the global scale, hydropower dams account for about 20 per cent of electricity generated, and in 24 countries, including Brazil, Democratic Republic of Congo,

Zambia and Norway, hydropower covers more than 90 per cent of national electricity supply needs. Half the world's large dams were built solely or mainly for irrigation. Between 12 per cent and 16 per cent of world food production is based on dams, and as reservoirs they provide protection against floods.

Unfortunately, this impressive balance is counteracted by comparably significant problems. Construction of large dams is a major intervention in the ecosystem of rivers and the lives of many people. The WCD estimates that some 40-80 million people, mostly indigenous peoples, have been displaced by reservoirs worldwide and robbed of their livelihoods from fishing or farming. Serious conflicts are simmering between neighbouring countries because dams have turned off the water supply for downstream states.

The late Indian Prime Minister Jawaharlal Nehru once said: "Dams are India's new temples". Right up to the 1970s, large dams were seen as the synonym for development and economic progress. Dam-building reached its peak between 1970 and 1980, when an average of two to three new large dams per day were commissioned. But a considerable number of the dams analysed by the WCD have fallen short of their technical and economic objectives. Construction cost overruns averaged 56 per cent. Many dams have had negative ecological impacts, and the disadvantages for people living downstream were mostly not taken into account. The planning of dams did not examine sufficiently possible alternatives for meeting power and water needs. There were hardly any retrospective evaluations of dam projects.

A New Framework for Decision-Making?

Despite this sobering stocktaking, the WCD arrives at an astonishingly simple finding: dams are primarily a means to an end. Their task is to improve the well-being of the people on a sustainable basis. This improvement should be economically acceptable, socially just and environmentally sound. If this goal can be achieved by a dam, its construction should be supported. Where alternative options offer a better solution they should be the prepared choice.

The Commission based its work on a set of five crore values for future decision-making: equity, efficiency, participatory decision-making, sustainability, and accountability. With regard to legal aspects and the extent of the potential risks for those involved, the WCD proposes development of an approach based on recognising rights and assessing risks. All risks-bearers should have a place at the negotiating table.

The WCD also recommends seven strategic priorities for decision-making: gaining public acceptance; comprehensive options assessment; reviewing existing dams; sustaining rivers and livelihoods; recognising entitlements and sharing benefits; ensuring compliance; and sharing rivers for peace, development and security. These priorities are reinforced by 23 practical criteria and guidelines which can be adopted, adapted and applied by all actors involved in the dam controversy. For instance, the WCD suggests analysing points at issue together with the people affected by existing dams and developing joint proposals for solutions. People affected by new dam projects should be among their favoured beneficiaries, and their claims should be made legally binding.

The report offers a comprehensive compilation of knowledge, which previously was limited to individual case studies or a narrow specialists contexts, on the social, economic, technological and ecological problems and impacts of large dams. The makes the report a central reference which helps greatly in bringing objectivity into the debate. Provision of an analytical framework and strategic options is certainly an important step. But with regard to its task of developing internationally valid criteria and guidelines for the planning, design, appraisal, building operation, monitoring and shutdown of dams, the report remains very general. What will be decisive here will be to practice with the relevant actors the method and content of the suggested mediation process on the basis of specific cases.

Deeds Must Follow Words

The WCD's work must now be made useable for the private sector, civil society and development purpose. What is required is a discussion process involving all major actors. The objectives of this process must be the development of practical and effective guidelines in addition to the current standards.

The WCD calls on bilateral development organisations and multilateral development banks to support only dam projects that have resulted from an open process of examining various options. The parties should observe the WCD guidelines. Measures to save water and power should be examined and, if applicable, be promoted.

Private sector companies should publish guidelines on corporate behaviour and acknowledge the WCD principles, criteria and guidelines. Further, the private sector should draw up and implement voluntary codes of conduct, management systems and certification procedures, such as the internationally recognised standard for environmental management (ISO 14001). The OECD's Anti-Corruption Agreement should be observed and declarations of honesty incorporated in contracts. Business associations should develop process to monitor compliance with the WCD guidelines.

NGOs should primarily check compliance with agreements and assist aggrieved parties to seek compensation. They should also assist in identifying relevant stakeholders for dam projects, using the rights and risks approach. Finally the NGOs should build up support networks and partnerships between them.

Importance of Information

But do these noble proposals provide the whole answer? The WCD process is based on the opportunities for personal development of every individual in an open society. But it is not enough to build on the negotiating abilities of the potentially affected alone because only specialists can anticipate the complex impacts of dams. Therefore participation presupposes that the mediation process contains a substantial informative component.

Relying solely on a mediation process is also not sufficient in providing for social impacts. There is no generally recognised method to determine the value of 'goods' in a subsistence economy. Without objective criteria, the moral demands on both sides are extremely high. Strategically-motivated behaviour will then not be prevented even if all participants agree readily that the subjective standard of living of people affected by a dam should be improved or at least maintained.

Mediation processes make sense only if agreement are observed. According to the WCD analyses, lack of compliance with agreements is the main cause of the negative social impacts of dams. In the case of dam projects co-financed by international donors, disbursements of funding instalments could depend upon independent evaluations that confined compliance. Ensuring compliance is much more difficult in the case of projects financed by the private sector. Because there is no independent institution that could assume the role of arbitrator, it must be in the business world's own interests to act responsibly in ecological and social terms. To be credible, it must provide transparency and independent certifiers.

Due to the opposing interests involved, no-one should expect reaching consensus to be easy. But the example set by the WCD is not the only reason for hope. Taking a closer look at it, the model offers significant advantages for all participants. Partner countries and development organisations wish to continue to use the potential for development which dams will also offer in the future. The private sector will also continue to build and operate dams, and for they need planning certainly. The advantages for NGOs and the people affected are also obvious.

29

Major Cyclones in Andhra Pradesh: Some Observations

Cyclone is considered as the most devastating of all natural phenomena. Repeat cyclones create no sense of security to the wealth and life of the people. It is also well-known that the sudden and rapidly developing cyclonic storm not only disrupts the prevailing order of life and produces danger, death and loss of property to large number of people residing within a geographical area but also creates environmental imbalances.

Andhra Pradesh is one of the states facing moderate to severe and repeated cyclones affecting the life of the people of the state and especially coastal districts. In the history of Andhra Pradesh, there is a long list of cyclones. The cyclone which hit the town of Machilipatnam in the year 1864 was one of the severest in the history of the world. The tidal wave was so vast and rushed with such force that it submerged the coastal areas deep upto twenty miles. It took many weeks for the flood water to sink. The entire area which was submerged under sea water become saltish and consequently infertile for many years to come. It appears that in the year 1796 also Michilipatnam and surrounding areas were hit with a big tidal wave which claimed over 20,000 people besides loss of property worth more than Rs. 250 crores then. The impact of cyclone of 1864 was more on life and property of the people. More than 35,000 people and 3 lakh livestock dead the worth of nearly Rs. 600 crores property was lost. Almost all the houses at Machilipatnam were completely ruined. Even the mighty fort of Britishers turned into shambles but for one building and the Bell tower of the church. With this calamity and Machilipatnam being prone to frequent cyclones, the Britishers shifted all their establishments to Madras, after this cyclone. The Britishers erected two monuments in memory of the 1864 cyclone victims one of Robertson Square in the heart of the town and the other at Machilipatnam Fort.

Relief measures were taken up to persons who were deprived of the essential needs of life because of natural disaster resulting from cyclone of 1864. Cyclone relief consists largely of emergency provisions of food, clothing, medical care and shelter with the aid of voluntary contribution from other communities and countries. Even International Read Cross an other voluntary organisations did not take cyclone relief as one of the chief activities. Only in 20th century these organisations have been extending assistance to the victims besides greater role of the central and state governments in relieving the population affected by the cyclones.

The second major cyclone in the state like that of 1864 cyclone, hit the coastal region on the night of 19th and 20th November, 1977. There was terrific gale with speed ranging 120-150 km. per hour. Coastal districts of Andhra Pradesh such as Krishna, Guntur, East Godavari, West Godavari and Prakasam and to a small extent Srikakulam, Nellore and Visakhapatnam districts were affected. The most affected areas were Divi, Machilipatnam, Bapatla, Repalle and to some extent in Chirala Taluk. Huge tidal waves engulfed the coastal region of the Divi, Machilipatnam and Repalle Taluks. There was extensively damage to life and property. Approximately 71 lakh persons in 2302 villages were affected by the loss of property, crop and other damage. 7932 persons lost their lives. In Krishna district alone 6706 persons died. The loss of cattle was 2,32,046 and 45,530 other livestock. There was substantial damage to houses. 86,650 huts were completely washed away. Approximately 2,16,000 persons have either lost their houses or have suffered other loss. There has been widespread damage to crops worth of Rs. 450 crores and substantial damage to public buildings such as roads, electrical installations, highways, railways, telephonic installations, irrigation canals etc.

To overcome the effect of 1977 cyclone, 172 relief camps were opened in Krishna, Guntur, East Godavari and Prakasam districts. About 2 lakh persons were provided shelter immediately. More than 25,000 quintals of rice and food packets were distributed. Medical special staff was sent to the affected areas to assist the Collectors to take measures against outbreak of epidemics. Financial assistance was given for house repairs and the Government sanctioned remission of land revenue in the most affected areas of Machilipatnam, Repalle, Bapatla and Divi taluks.

Andhra Pradesh again experienced a major cyclone disaster in May 1979. Besides loss of life there was enormous damage to public and private properties in coastal districts. Prakasam, Nellore and Kurnool districts were the worst hit. The heavy rain under the influence of the cyclone, affected the districts of Guntur, Krishna, West and East Godavari on the coast line and Kurnool and Cuddapah districts of Rayalaseema and Mahaboobnagar district of Telengana. This was most unusual occurring in the month of May and again touching all the three regions in the state. Due to precautionary measures like providing information in time to the people of the state evacuating all the people to safer places mobilizing and gearing up of administrative set up to take immediate steps, death toll and loss of property were greatly reduced though the intensity of the cyclone was double than that of the one which occurred in 1977.

Nearly Rs. 60 crores worth of crops were damaged in the State. The crops in Nellore district were heavily damaged followed by East Godavari, Guntur, Cuddapah and Prakasam districts. Due to intensive precautionary measures during precyclone period, the death toll was drastically kept under limit with just 750 human lives though the severity of it was very high compared to the earlier cyclones. About 3 lakh livestock were lost due to improper care taken during per-cyclone and during cyclone periods. Out of the six cyclone affected districts, Prakasam lost heavily its livestock. Most of the deaths in this district were due to surging waters which breached the tanks with extensive loss of livestock and damage to the property.

More than cyclone relief measures, 1979 cyclone reveals that the success of the excellent preparedness measures taken by the State could reduce the death toll greatly. Wholesale evacuation of the people living in low lying areas in coastal areas prevented not only the number of deaths and loss of property but also could reduce the risk of undertaking relief measures.

Another severe cyclone was experienced by the people of Andhra Pradesh in October, 1983. As Andhra Pradesh is called the rice bowl of India and more especially East and West Godavari districts, paddy crop was completely damaged. All the major rice producing districts such as East Godavari, West Godavari, Nizamabad and Karimnagar districts were severely affected besides Visakhapatnam, Khammam, Warrangal and other districts, though the loss of human life was below 200 the cyclone could affect 50 lakh people in 6,322 villages of 14 districts. More than 2 lakh houses were completely damaged and nearly one lakh and 50 thousand houses were affected partially 12,000 cattle and 31,000 livestock perished in the cyclone. Standing crops in 1,38,500 hectares were completely lost and crops in about 31,13,150 hectares were partially damaged in addition to dry crops in 3 lakh hectares of land. The loss of both public and private properties was estimated around Rs. 600 crores. The road length of more than 8,500 kilometers damaged completely. This was considered as one of the most severe cyclones in recent years as it could affect the economic structure of the State very badly.

Like that of 1979 cyclone, with utmost pre-cyclone preparation and protection, death toll and loss of property were minimised to the lowest possible level. Thanks to the immediate steps taken by Telugu Desam Government and officials incharge in the districts, a major threat was averted with minimum loss by evacuating the people to safer places in almost all the coastal districts. Indian Army and Navy, Police and other officials could evacuate more than 50,000 persons in Visakhapatnam, East and

West Godavari, Guntur, Krishna, Nizamabad, Karimnagar and Nalgonda districts to safer places. Hon'ble Chief Minister, Sri N.T. Rama Rao, has sanctioned immediately Rs. 30 crores towards relief without waiting for Central assistance.

After five months, in February 1984, the districts of Nellore and Chittor have again affected by a cyclone. The most worse unusual cyclone in the month of February could bring misery to farmers in more than 150 villages by destroying standing paddy crop and gardens worth of more than Rs. 20 crores and loss of property in the form of collapse of houses worth more than 50 crores. It is, however, the administration of organised emergency relief and rehabilitation could set right the imbalance. The cyclone of 1996 has the same impact on A.P. economy.

The cyclones which affected the people of Andhra Pradesh demonstrate that the people affected by cyclone disaster are not confined to the immediate geographic areas of who identify themselves with, persons and organisations in the stricken area are also affected. Thus the effectiveness of cyclone relief measures are usually national in scope. In each cyclone period, effective preparations were made to face cyclone with necessary arrangements by evacuating the population to safe places.

But there remains a problem of co-ordination and control in undertaking relief operations. Shortly after the cyclone impact, thousands of persons have to converge on the affected area and on first-aid stations, hospitals, relief centres and communication centres near the area. Along with this movement of persons, incoming messages of anxious inquiry and offers of help from all parts of the country and even from foreign countries should come forward to set right communication facilities and to supply food, clothing, bedding and other materials. This action should continue for a few weeks following a cyclone. At times of cyclone generally there would be confusion due to lack of systematic procedures for maintaining a central strategic overview of the cyclone to apply controls and resources where they are most critically needed. Communication is often inadequate partly because of the destruction of communication facilities but more generally because of improper use of these facilities. With greater coordination and cooperation from the people and making them to feel a great sense of urgency to help the victims, these effects of the cyclones would be greatly minimized in future.

30

Energy

It has been scarcely 200 years—the dawn of the Industrial Revolution—since humans abandoned sole reliance on firewood, other biomass fuels, and direct sunlight to meet daily energy needs. In the past half-century, global demand for energy grew twice as fast as population, as industrial nations burned coal, oil, and natural gas to fuel their economies. Over the next half-century, world energy demands are projected to continue expanding beyond population growth, as developing countries try to catch up with industrial nations.

Developing countries will see tremendous growth in energy consumption in the next half-century, as growing populations and increasing affluence combine to drive their energy demands to dizzying levels. Based on projections from the U.S. Department of Energy and the Intergovernmental Panel on climate change, total energy consumption in the developing world will grow by 336 per cent—nearly three times faster than population—over the next 50 years, from 3,499 million tons of oil equivalent to 15,255 million tons. By 2030, energy consumption in the developing world will likely surpass usage in industrial nations.

Rising per capita consumption accounts for nearly two thirds of the growth in energy demand in poorer nations, but different population trajectories can have dramatic effects on future demands. For example, assuming the same growth in per capita energy demand, moving to the low U.N. population projection will reduce total energy demands from developing countries by 2,792 million tons of oil equivalent—the output of nearly 3,000 average-sized coal-fired power plants.

In the next 50 years, the greatest growth in energy demands will come where economic activity is projected to be highest: In Asia, where consumption is expected to grow 361 per cent, though population will grow by just 50 per cent. Energy consumption in Latin America and Africa is projected to increase by 340 per cent and 326 per cent, respectively. Lower rates of population growth in Asia, compared with Latin America and Africa, mean that energy use per person will increase most in Asia. Nonetheless, in all three regions, local pressures on energy sources, ranging from forests to fossil fuel reserves to waterways, will be significant.

When per capita energy consumption is high, even a low rate of population growth can have significant effect on total energy demand. In the United States, for example, where current per capita energy demand is nearly double that in other industrial nations and over 13 times that in developing countries, the 75 million people projected to be added in the next 50 years will boost energy demands by 758 million tons of oil equivalent—roughly the same as the present energy consumption of Africa and Latin America.

World energy use per person doubled between 1950 and 1973, before confronting a short-term slowdown when restricted exports from oil-producing nations drove up energy prices. Another price shock, combined with a global economic recession, resulted in the slowdown of the early 1980s. The most recent stumbling block in energy growth followed the 1989 revolution in Eastern Europe, when energy use in the former Soviet states plummeted. Although DOE and IPCC project substantial future growth, similar forces may act to check such a development.

World oil production per person reached a high in 1979 and has since declined 23 per cent. Moreover, estimates of when global oil production will peak range from 2011 by petroconsultants to 2025 by the IPCC, signaling future price shock as long as oil remains to world's dominant fuel. Although people born in 1950 saw per capita oil production quickly double in a few short decades, those born in 2000 are likely to see it cut in half, dropping below 1950 levels.

In addition, meeting increased energy demands will require more storage and transportation infrastructure. Communities without a reliable supply of clean water or an adequate system for waste disposal may also fall short in connection to power supplies. For the estimated 2 billion who are still off the grid—and also experiencing high rates of population growth—decentralized energy technologies, such as solar roof shingles and fuel cell power generators, are likely the most feasible and affordable option for meeting increased energy demands.

Yet, it will not necessarily be the scarcity of fuel that constrains future growth in energy consumption, but rather concerns about climate change, air quality, and water quality. Growing climate concerns will require massive reductions in fossil fuel use at a time when demand for energy is soaring. A shift to renewable energy sources, such as solar energy and wind power, in addition to continued efficiency gains for power plants, cars, and appliances, holds great promise for meeting future energy demands without adverse ecological consequences.

31

Energy: A Fair Deal for All

Both the supply of energy and the demand for it have spiralled in modern societies, where everyday life and changes to the environment, global as well as local, are conditioned by energy production and use. There is a crying need for a fairer share-out of material goods, energy and economic resources.

Energy comes in three forms: So-called 'fossil' fuels (coal, oil and natural gas); nuclear power; and 'renewable' energies (hydroelectric power, thermal or photovoltaic solar energy, wind and tide power, wood, etc). Each of these has its own undeniable advantages and drawbacks.

Fossil Fuels

Fossil fuels are abundant and very simple to use. Oil, for example, can be very easily transported and processed, and is relatively cheap. The technology for producing its many derivatives is highly developed. What's more, it is particularly well suited for use in all forms of land, sea and air transport. Its handy fluid form and its price make it appropriate to the needs of poor communities or those that are unable to invest in capital goods.

Fossil fuels account at present for 77 per cent of all the energy produced and will, according to the most realistic projections, still account for 73 per cent in 2020. The resources will be strictly limited geographically as well as in duration, being restricted to certain regions. This state of affairs is fraught with the risk of tensions and even conflicts, owing to the strategic importance of energy supplies.

Fossil fuels, are furthermore, responsible for the man-made increase in the carbon dioxide content of the earth's atmosphere, with the associated danger of an increase in the greenhouse effect and, as a direct result, global warming of the order of 1° to 4°C in the next twenty years, which would adversely affect the climate and the environment. Though much uncertainty remains as to the scale of these effects, the risk is great enough to mean that every effort should be made to slow down the increasing 'carbonization' of the atmosphere due to the intensive use of fossil fuels.

Nuclear Power

The main advantage of nuclear power is that it has no effect on the carbon dioxide content of the atmosphere. As it is also cheaper (per energy unit) than hydroelectric or thermal energy, some countries, have adopted strongly for this way of producing electricity.

Nuclear power is, however, far from being unanimously accepted. Public opinion is very conscious of the lack of candid information and of the safety of nuclear plants, two aspects that have not always been treated, in some countries, with all the necessary care and clarity by the authorities and the operators. The public is also worried about the disposal of long-lasting radioactive wastes, an acute problem to which the experts seem confident that a long-term solution can be found. It would also be a mistake to underestimate the danger of the spread of nuclear arms, even though the main powers are now significantly reducing their arsenals of these weapons. A final point is that only those countries which can afford to make the huge investments required can put nuclear plants into operation. The investment is offset by the low cost of the fuel but is recouped only in the medium and long term.

Renewable Energy Sources

The ecological movements, which are worried both by global warming and by the real or imagined dangers of nuclear power, would like renewable energy sources to be developed faster than is now the case. These forms of energy at present supply some 18 per cent of total demand, which puts them well ahead of nuclear power.

Technology is moving rapidly forward in this field. These forms of energy are capable of meeting the needs of communities that it would be too expensive to connect to a central grid supply, but despite improved productivity and falling costs, they remain on the whole dearer than the two previous forms. It will be a long time before they can constitute the main source of supply. Other problems that remain to be solved include the major investments required for hydroelectric power stations and the environmental damage caused by the building of dams and wind farms.

We must face the fact that as of now there is no 'miracle' energy that is risk-free for humans and their environment and is also cheap and inexhaustible. There is no such thing as absolute security as regards power generalisation and use, and it will not be possible to the future to do without any of the above-mentioned sources. Energy demand will continue to grow as a result of irreversible technological advances, of the justified demands of the non-industrialized countries, and of population growth that is in any case set to continue for at least the next fifty years.

Some Ethical Principles

A number of imperatives must thus be borne in mind by every individual, every nation and, in particular, the citizens of the industrialized countries. These are: the right of each individual to sufficient sources of energy; our responsibility towards our children and our children's children; protection of the environment; prevention of the potential major risks from the production of energy on a massive scale; the control of costs and the need to carry on with research in all these fields.

Some of these obligations—those relating to population growth, climate change or the disposal of nuclear wastes, for example—are of a very long-term nature, while others-efforts to deal with pollution caused by road transport or chemical waste disposal—are short-term. These differences of time-scale and the serious possible interactions between the quantitative and qualitative aspects of the question have to be taken into account in observations of an ethical character such as the following:

- The present situation, whereby nearly one person in four in the world is without access to the energy resources he or she requires, cannot be accepted with registration. Those with an active role in energy policy-decision-makers, industrialists, research workers and so forth-must ultimately ensure that their exist, and continue to exist, sufficient resources of sufficiently cheap energy for all countries to have access to them, regardless of their geographical or economic situation;

- There should be no pretext for unnecessarily keeping the countries of the South, which urgently need proper infrastructures, on short commons as regards energy use. This is one area where, more than in any other, people need to be informed, so that, they can take part in discussion and decision-making on subjects where scientific and technological knowledge is essential;

- Our duty to future generations enjoins us to use energy resources as sparingly and rationally as possible, especially as we know that a major part of these resources may be exhausted in a century or two;

- Even though rapid progress is being made in the exploration of space, we must acknowledge the obvious fact that we have only one Earth and must therefore preserve and protect it. Since energy production and use may jeopardize our environment, there is an urgent need for appropriate measures to be taken as rapidly and as effectively as possible. The management of nuclear waste and campaigns to combat all forms of pollution arising from energy use constitute unconditional obligations in this connection;

- Whenever massive quantities of nuclear or other forms of energy are produced or transported, *e.g.* when oil is transported by sea or big dams are built, major risks to life and health ensue. Absolute safety is unattainable, but the various energy authorities are nevertheless under an obligation to issue and enforce appropriate safety regulations;

- Unit cost will continue to be the main factor influencing the choice between different forms of energy. Production costs must be controlled and savings constantly sought if energy supplies are to be available to all;

- Research sometimes seems to have been neglected in work on energy production and consumption, but it is an indispensable duty. Efforts to find new sources of energy and more economical ways of using it must continue.

32

Energy and Sustainability

Mankind's history is marked by a growing use of energy which until the end of the Industrial Revolution came largely from renewable sources. It was coal that fed the furnaces and boilers of the Industrial Revolution from the end of the seventeenth century to the nineteenth century, and drove railway transport and steamships. As well as being a useful source of mechanical energy, it was also used in the manufacture of coal gas for street lighting and in the chemical industry. In fact, coal was the principal form of energy until 1900.

Discoveries at the beginning of the nineteenth century allowed the use of electricity and revealed the relations and the interconvertibility of different forms of energy. The principles of conservation and of energy quality did not become operative until much later. Meanwhile, in 1882, the first system for producing and distributing electricity in a large city was installed. This was the beginning of the second phase in industrialization through electrification.

Following the first successful oil drillings in 1859, Standard Oil, the first of the modern large scale oil companies, attempted the first vertical structure for overall control of the oil process. It involved extraction from the subsoil, storage, refining and final distribution. Later, the growth of derivatives, the lower extraction costs compared to coal and the greater ease and economy of transport made oil modern society's basic energy source.

The internal combustion engine led to motorization on a massive scale by land, sea and air and guaranteed a constantly growing market for petrol. The forties marked the start of the new petrochemical industry, which gave rise to an enormous number of new products; synthetic rubber, plastic, medicines, cosmetics, artificial fibres, detergents, weedkillers, fertilizers, butane, propane, etc., opening the way to the mass-production of consumer goods and introducing new, non-biodegradable substances into the environment.

After World War II, ambitious programmes to produce electricity from nuclear energy were begun, in the search for a return on the enormous amounts of money invested. The economic expansion in the West during the fifties and sixties was directly related to enormous petrol consumption at a time when energy was considered plentiful and cheap. Energy consumption during these decades grew more than exponentially. The fastest developing industrial sectors were precisely the ones that consumed most energy—petro-chemical industries, metallurgy, car manufacturing, domestic appliances, electricity generating, etc.— and a trend developed towards goods and services with higher energy intensity. Since 1950, increased energy production has been systematically favoured over more rational use. So much so that the increase in energy consumption has been taken as a reliable indicator of progress.

The Aftermath of the Oil Boom

The oil crises of 1973 and 1980 showed up the fragility of an energy system that was over-dependent on oil. The War in the Gulf was reminder of what was at stake for the Western economies; free access to cheap oil in the Middle East. It was therefore fear of the hardship caused by the first crisis that brought about a change in attitudes in Western countries; efforts were directed at breaking free from this dependence, diversifying sources, perfecting, replacement energies and promoting energy-saving programmes.

The eighties marked a change in people's awareness about environmental problems. The damage was making itself felt in more and more places and eventually the global threat to our planet as a result of our energy system became clear; the composition of the atmosphere was changing and could lead to possible changes in the climate.

According to recent figures, 82 per cent of all the energy consumed in the world is produced by burning fossil fuels, 7.5 per cent from burning biomass, 5.5 per cent from the use of hydraulic energy and 5 per cent from nuclear energy. Most of our energy in other words, in non-renewable; it runs out as we use it, as the population increases; and it comes from fossil fuels, which on burning increase the amount of CO_2 in the atmosphere. If we add to this the accumulation of nuclear waste, the problems of access to oil deposits, constant spillages during transport and all the different imbalances involved in the world energy system, the outlook is far from sustainable.

The inequalities speak for themselves; globally, less than a quarter of the world's richest population consumes almost three quarters of the energy commercialized in the world. For example, the average annual consumption per capita in the United States is 26 times higher than in India.

The Choice of Change

Opening the way to societies that make sustainable use of energy necessarily involves increasing and improving energy efficiency, both in supply technologies and in end-use technologies, at the same time using renewable energy sources instead of fossil fuels.

Choosing the right system for the transformation of primary energy sources into energy services such as lighting, cooling, cooking, mechanical force, transport, etc. and choosing the most suitable appliances and technologies in each case is fundamental.

The truth is that a good standard of living is possible without wasting anything like as much energy. A series of relatively straightforward measures today allow a far higher level of comfort than in 1950, using one third as much energy for heating water for washing in the home.

Petrol consumption by vehicles has dropped by 40 per cent in forty years, from 8 litres/100 kilometers to 5.3 litres in some models, and the work of improving their energy efficiency continues. In industry, the energy consumption necessary for manufacturing large intermediary products (steel, cement, paper or fertilizer) is decreasing steadily at a rate which varies between 0.5 per cent and 0.2 per cent per year according to the product.

Today's incandescent bulbs consume one twentieth as much electricity as bulbs in the twenties. The compact fluorescent bulbs now available can cut this down again to one fifth. Efficiency in lighting has increased one-hundredfold. The use of new materials and a more rational use of traditional materials allows a reduction in the amount of energy and raw materials consumed. Building a house, for example, requires 20 per cent less energy than in 1950; building a vehicle, 40 per cent less. On a global level, reducing our security's energy-intensiveness is the first step towards energy sustainability.

33

Turning on the Heat—India's National Programme on Solar Cooking

Anyone who has ever watched a pot of water being boiled just by the sun is sure to have been impressed. High-quality solar cookers can reach temperatures over 200ºC. This temperature is more than sufficient to cook food, bake bread and heat up an iron for ironing clothes. As rural people in many parts of the world heavily depend on wood for preparing meals, solar cooking can also make an important contribution to saving our forests. But even after three decades of implementation, solar cookers have not evolved into commercially viable products that sell themselves. India is the country with the most experience in this field.

The principle behind solar cooking is as fascinating as it is simple: sun rays are converted to heat and conducted into the cooking pot. Solar cooking also has some practical advantages: solar power is inexhaustible, clean and free. By lessening their dependency on conventional fuels people can save money and non-renewable resources as well as the environment. The solar cookers often make use of a box or parabolic reflector to concentrate the sun's rays.

In India some 4,75,000 solar cookers were sold last year with the help of government grants and highly subsidised prices. That puts India way ahead of all other countries. In rural homes—70 per cent of the Indian population lives in rural areas—cooking accounts for a major share of total energy consumption. The available energy sources are firewood, crop residues and animal dung:. In urban and semi-urban areas, gas, kerosene oil and coal are used for cooking purposes. The smoke emitted from these fuels pollutes the environment and the kitchen, affecting the health of the family members, especially the women. Fuel wood is becoming scarce due to the depletion of forest.

Solar Energy is Abundant

Solar cooking has long been envisaged as a solution to mitigate these problems. After all, solar energy is abundantly available in most parts of the country. The daily average solar energy incidence ranges between five to seven kmwhr/m^2 and there are as many as 250 to 300 clear sunny days a year. On such days it is possible to cook both mid-day and evening meals in a solar cooker. It is, however, recognised that solar cooking cannot fully replace conventional fuels. The Government of India first initiated efforts in the early 1980s to popularise solar cooking devices all over the country in order to reduce dependency on conventional fuels.

National Programme Launched

A national programme on solar cooking was launched during 1981-82, which promoted above all the box-type cooker on account of its relative advantages. The concentrating types were not selected due to their high cost, the need for frequent tracking and the fast deterioration of reflecting surface etc. A subsidy of 33 per cent of the cost was granted by Central Government. In some states, an additional subsidy was also provided. To maintain the quality of the solar cookers being marketed, standard specifications were developed and given to all manufacturers and state agencies. Extensive efforts were also made to promote a wide network of manufacturers so that, over the years, around 55 manufacturers established units. The programme driven by the subsidy scheme was continued until

1994, when it was decided to adopt a market orientation and introduce new arrangements.

Studies to monitor practical benefits were conducted and revealed that while over 70 per cent of the owners were using their cookers, many did not use them frequently. The rest were not cooking with them at all due to climate problems, non-availability of open space, time constraints or cookers being out of order. Main requirements in the field of service were found to be the black-painting of tray and vessels, removal of moisture between glazings, replacement of broken glasses and prevention of hot air leakage. Shortcoming of the programme itself also became clear. As the manufacturers delivered the cookers to the state agencies who were selling them on they were mainly interested in supplying cookers and gave insufficient attention to after-sale service for the product. They were not interested in developing their own sales and services network. This meant that customers could not get a model of their choice. All these factors finally led to the termination of the subsidy regime.

New Commercialisation Strategy

In order to give the programme a stronger market orientation, a new strategy was evolved in 1994 aimed at the commercialisation of solar cookers. Under this scheme, manufacturers were allowed to make modifications to the cooker design to make them more attractive and user-friendly. Sales were allowed both through state agencies and directly by manufacturers through their own network.

Certain negative trends have also emerged: Contrary to the original intention, it has become apparent that most solar cookers were being sold in cities and suburban areas. A price of 25 to 60 US dollars—depending on the model—is still too high for the rural population. Nevertheless, the programme focuses on commercial success in view of the excellent progress now being made. The market-oriented programme is leading to the creation of independent distribution networks, better, customer service and user-friendly models. However, commercialisation has, for the time being at least, brought higher prices and falling output.

"Low-cost but durable models still need to be developed for wider dissemination of the technology in rural areas.

Sales Showrooms

A new initiative by the government is focusing on the establishment of showrooms for the sale and servicing of renewable energy products (including solar cookers) in major cities of the country. With financial help from the government they will offer over-the-counter sales of different renewable energy products, disseminate information and carry out repair and servicing of the equipment. These solar shops, which operate under the name of 'Aditya' (means Sun in Sanskrit), allow manufacturers from all over the country to present their solar cookers and other products to the public. The average sale during the last five years numbered 25,000.

To boost the sale of solar cookers and provide after sale services to users, a new scheme for self employed workers (SEWs) has been introduced in 1999. Under this scheme, state agencies enlist a set of SEWs in their respective states who will receive training in the repair of solar cookers and in their proper use and preparation for various types of cooking. The SEWs will be recruited among unemployed youths with a technical qualification and experience in mechanical work. The scheme worker is to act as a link between the users and the manufacturers and banks. The SEW shall receive a nominal payment per cooker as a promotional incentive for selling more cookers in his area.

Never were these cookers more necessary than today. Wood burning still meets 15 to 18 per cent of primary energy consumption—more that nuclear and hydro power together. Greater demand for fuelwood is not only leading to desertification, it will also leave a growing number of people (the FAO puts the figure already at two billion) without enough energy to prepare themselves a regular hot meal.

34

Between Wish and Reality

The Limited Potential of Solar Cookers

Efforts to promote the idea of using solar cookers have been made for decades, but despite great commitment they have failed. Nowhere has it come to a self-sustaining commercial dissemination of these cookers. Above all, it has not been possible to reach; the main target group, the rural population. But that has not detracted from the fascination of using the sun to cook.

The Indian Experience

Most experience with solar cookers has been gathered in India. The National Physical laboratory in New Delhi tested the efficiency of cooking boxes and concentrating cookers back in 1953. Large batches were produced at the beginning of the 1960s, but evidently without lasting success. That meant when in 1981, a national programme to disseminate cooking boxes was launched it had to begin all over again. Over the course of time and with strong government promotion, 55 manufacturers began producing the units. Subsidies pushed the sales price down to half the production cost of US$ 60-70, and in low-income areas even down to US$ 15.

At the same time, the government ran a radio, television and newspaper advertising campaign to promote the cookers, and roped in prominent figures to endorse them. Other promotions featured cooking demonstrations in villages, and giving a solar cooker as a prize for couples that danced best at parties. About 1,20,000 cookers were sold in this way by 1990. But that was only a deceptive success. Random checks found that after a while most households used their solar cookers only occasionally. About 30 per cent no longer used them at all for various reasons.

The results were especially poor in the villages. A survey of rural energy consumption in six Indian federal states in 1966 found that of 51,000 households only 70 possessed a solar cooker. At the same time, the survey also refuted the oft-repeated prejudice that the people opposed innovations and rejected them even when they were useful. After all, 6,200 of the households surveyed had brought a pressure cooker because they cooked faster and used less fuel.

India is now making a fresh attempt to propagate solar cookers. According to the Ministry of Non-conventional Energy Sources, almost 500,000 highly-subsidized cookers were sold in 1998. It is not clear how better results are to be achieved this time. The programme director is surprised that most of the cookers are sold in town rather than in rural areas, although based on earlier experience nothing else was to be expected.

Socio-Cultural Factors

The causes of the only modest success of solar cooking equipment lie less in the technical sector than in the socio-cultural, socio-economic and psycho-social area. This profound-sounding formulation identifies what is basically a simple fact. Solar cookers function, but they cannot replace the customary cooker. Even in arid zones the sun does not always shine, and in the mornings and evenings cooking must anyway be done by customary methods. That means solar cookers are a supplementary way of cooking that saves fuel. Experience has shown, however, that it saves at most one-third of usual fuel consumption. That is not fundamental progress for a household, only a degree of improvement. It is quite different from the situation in, for example, the photovoltaic (PV) sector, which enables completely new things such as electric light and radio reception. So it is not surprising that small, portable PV systems are very much in demand even among Tibetan nomads.

By contrast, a solar cooker is rated by purely economic criteria. Savings of time or money are balanced against the cost of the unit and its limits. Not all meals can be cooked on a solar cooker, preparation takes longer, and the cooking process can no longer be controlled as one like because it is determined by the amount of energy the sun offers. These disadvantages are not due to any particular shortcomings of the solar cooker as such which could be remedied by further development, but to the vagaries of the sun itself as a source of energy.

Tibet as a Special Case

The only region in the world where solar cookers are firmly established is Tibet. A total of 70,000 to 1,00,000 units are estimated to be in use among a population of two million. But Tibet is a special case in which several favourable conditions for hours of sunshine per year, or an average of eight hours of sunshine per day, and the intensity of its solar radiation is so high that with only a few exceptions solar cooking is possible throughout the whole year. Since sunshine in Tibet means almost always direct radiation, concentrators with parabolic reflectors are used.

On the other hand, Tibet has an extreme shortage of other fuels. Firewood is simple not available. There are hardly any trees in the mountains. Other traditional sources of fuel such as shrub, yak dung or turf are also extremely scarce and hardly available for the inhabitants of the three cities of Lhasa, Shigatse and Gyantse. They are dependent on cylinder gas or kerosene that must be brought in over thousands of kilometers from China and is correspondingly expensive.

Local eating habits also favour the use of solar cookers. For the yak butter, tea Tibetans drink incessantly throughout the day a household needs a continual supply of hot water. The concentrator cookers, which can boil five liters of water in 15-20 minutes, are highly suitable for that. They are also used to cook rice, but very little for other meals. The structure of Tibetan housing areas is another factor favouring solar cookers, their enclosed courtyards and flat roofs providing, plenty of space for setting them up. All that has resulted in almost every urban household possessing a solar cooker. But use of the cookers in rural areas in much less widespread because the people have less money and transporting the heavy units to villages is difficult and expensive.

Regional Contribution

Favourable conditions like those in Tibet are seldom to be found anywhere else. Viewed realistically, solar cookers are not the global solution to the firewood crisis. At best, they make a regional contribution. Portraying the dissemination of solar cookers as a benedictory concept also does not help it further. Most of the solar cooker enthusiasts are still far from this insight. The World Solar Cooking and food Processing Conference, held in Varese, Italy, last October and attended by 300 experts from 64 countries, once again provided abundant material in support of solar cookers. The conference's final statement said: "Solar cooking has the potential to be one of the most significant contributions to solving the firewood problem".

35

Employment and Poverty Alleviation

Today the key socio-economic problem is large-scale unemployment. Spreading joblessness brings many other problems in its wake. It erodes national incomes and living standards, aggravating the already grindingly difficult job of promoting development and alleviating poverty. Joblessness also raises government budget deficits, increasing, macro-economic instability while soaking up investment for productive capital expenditure, education, training and relief aid. And joblessness ruins lives and communities by depriving people of the dignity and satisfaction that comes with earning one's keep and making a contribution to the well being of family and society.

Theories about how best to nurture development (and thus create jobs) have shifted considerably over the last decade. The state role has evolved, in the minds of many, from being a source of relief for the problems of unemployment, poverty and underdevelopment, to being a fundamental cause of these problems through the distorting impact of its intervention on the market.

However, the more market oriented philosophy that grew up during the 1990s has yet to provide convincing solutions in practice at least not on a grand scale, and especially not in terms of job creation as the present jobless economic recovery demonstrates.

The weakness of the current recovery and past approaches to economic development can be traced to the failure to consider employment as the predominant means of promoting growth and alleviating poverty. In policy circles it has too long been an almost ignored priority.

Current trends thus bode poorly, particularly as unemployment rates soar. In light of the circumstances, we need to begin re-examining some of the fundamental questions—if only to find out what has gone wrong with the answers.

Minimum Wage?

Let's begin with wages. With corporate restructuring in full force on a global scale, are low wage rates required to raise employment and maximize profits? A top manager of a multinational consumer electronics group certainly thinks so; he linked the perfect factory to a ship "so that we could move it around the world to where labour was cheapest". Perhaps, but this bottom-line emphasis on unit labour costs ignores at least two other factors; namely, that higher wages can act as a screen to select more productive workers and that higher wages translate into better productivity via improved worker nutrition, increased consumption and a generally healthier quality of life.

If higher wages bring these benefits (and it is an open question) should government insist that there be a minimum wage rate? Neo-classical economists tend to respond 'no', assuming that a higher wage rate puts money into the pockets of some low wage workers while forcing many others out of work because companies cannot afford to pay them.

Technology Transfer

The impact of technology is another area in need of study. Technological innovation is usually labour-saving and tends to originate in industrialized countries, moving toward developing countries like India, Pakistan where labour tends to be low cost and abundant. Would it therefore makes sense to slow down or somehow restrict technology transfer, especially to development markets, in the interest of preserving employment?

The answer here is clearly—no. Historical evidence abundantly demonstrates that attempts to retard technological progress bring about greater poverty and lower growth. Technology, infact, is at the heart of the new endogenous growth theory which is very much in vogue among development economists today. Slowing down or inhibiting technology transfer would certainly dash many countries' development hopes and aggravate poverty. However, the relationship between technology, development, employment and poverty alleviation is not without its complications.

In the 1980s, the buzz word among development specialists was 'appropriate technology', *i.e.*, small-scale and labour-intensive technologies that would increase productive output while allowing an equilibrium solution to be found such that ratio of the productivity of labour to that of capital is proportional to their relative prices. The conditions for this 'small is beautiful' approach to technology tended to be best met in agricultural production. However, where manufacturing industry is concerned, the small-is-beautiful approach foundered badly when the only viable technological alternatives proved to be highly capital-intensive.

Development Gap

A wide gap has emerged between developing countries with an inward focus (which tended to be protectionist and pursue policies of import substitution) and those with an outward focus and a policy of pursuing export-led growth. Competing in international markets requires technology that is as good as or better than that found in advanced, industrialized nations. Small, therefore, is not beautiful in the global manufacturing economy where product standards are high and the elasticity of substitution between labour and capital is very limited.

The drive to obtain state-of-the-art technology thus leads to a policy conundrum: it is a pre condition for success in manufactured exports, but the impulse to compete successfully in this most lucrative sector speeds up the transfer of technology from the developed to the developing world, thus reinforcing the bias toward labour saving equipment in developing countries and accelerating a process that is seen as a source of job loss in the industrialized countries.

Technology and Jobs

Before concluding that modern technology transfer is inimical to employment in developing countries, we have to distinguish clearly between technology's static and dynamic consequences. In a static sense, it is true that highly capital-intensive export industries may not create much employment on a net basis, but the dynamic effects of technology transfer do contribute to economic growth. And growth, in turn, generates multiplier effects in the form of demand, which stimulates ancillary production activities (like food processing or consumer goods) that rely on more labour-intensive technologies.

The problem is that the diffusion and application of technology on a global scale blurs the categories of international product specialization and creates a much more competitive and conflict-prone international environment.

For example, we have already seen the Asian Tigers move from producing goods such as textiles and processed food to producing hi-tech and value-added consumer durables. This advance is only possible due to the growth of human capital (facilitated by investment and higher incomes) and it leaves production of textiles to other industrializing countries, like Indonesia, the Philippines and now China. But the dynamic comes at the expense of jobs in industrialized regions, like the US and the EC, which lost more than a quarter of their work force in textiles during the 1980s. Inspite of job losses, advanced countries continue to produce textiles, notwithstanding major differences in the hourly wage rates for spinning and weaving and the fact that essentially the same hi tech equipment is being used in most production centres.

Protectionism

What has happened in textiles is happening in other industrial sectors (automobiles, for example) as well. The intense market competition is proving to be a source of trade conflicts, and possibly protectionism, as jobs come under increasing pressure.

For many workers and managers, the benefits of foreign direct investment look increasingly like a zero-sum game for employment, and there is a real risk that the tenuous link between overall growth and employment will break down altogether. It is hardly surprising that we are already seeing negatively affected workers and local businesses clamouring for protection in advanced countries.

Governments Role

The concerned governments are suppose to carry out much of this research. The three initial lines of inquiry follow from three reasonable assumptions about the future.

- First, increase in welfare and consumption subsidies are out; investments in training and human capital are in. How can investments in human capital be directed to positive employment effects? It is perhaps not time to explore more fully benefit schemes targeting the unemployed and the unskilled poor providing them with the type of subsidies that would enhance their human capital, improved their health and productivity through better nutrition and preventive medicine, and restore the dignity of holding a job?
- Second, given the quasi-inevitability of increased automation in manufacturing, how can other sectors (particularly agriculture and services) be developed to export their long-term potential for employment creation?
- Third, given the inevitable pressures of work and productivity in the global economy, what sort of alternative institutional arrangements need to evolve with respect to industrial relations, employment and work conditions?

Finding answers to these and other questions will require no small amount of new thinking, but parochialism or a failure of imagination would be fatal flaws in this global era.

36

What's Driving Migration

The scale and diversity of today's migrations are beyond any previous experience. Rapid urban growth and environmental degradation in rural areas have led to internal migration affecting hundreds of millions of people. Migration is now seen as a priority issue equal in political weight to other major global challenges such as the environment, population growth and economic imbalances between regions.

Families and households form the basis for economic growth, social development and personal fulfillment. Decisions, by individual women and men on marriage, family, a place to live, shape the destinies of communities and nations. National policies and international conditions provide the context for individual decision-making. Effective development policies, including population, reproductive health and family planning policies, address this reality.

Data on national and global population trends set the agenda for national policy. An important element of population programmes is gathering data that will allow policy-making, responsive to the realities of daily life, and to the needs and aspirations of individuals.

The dominant feature of global demographics is still growth. Age distribution is a growing concern, as the numbers of young and elderly people, grow, relative to the working-age population. The world is growing steadily more urban. From being a sign of strength and dynamism in the national economy, the rate and scale of urban growth has become increasingly a cause for concern. The influx of migrants to the biggest cities may be weakening both urban and rural sectors.

International migration is small in extent compared with internal movements, but has a disproportionate impact. Both internal and international migration are driven by population growth, and by inequities between countries. Migration is one of the choice which shape people's lives and the density of nations. But it can also be a symptom of inequity and underdevelopment. Migrants are by definition the most vulnerable members of the host community. Their living and working conditions should be protected.

Open and frank exchange of information and views between host and sending countries is needed more than ever. The aim of the international community should be to protect the right to move, but to ensure that movement is voluntary and that it stimulates rather than holds back personal and national development. "The point of departure should be the human right to live and work where one pleases, so long as it does not infringe on other people's rights to do the same".

The Urban Transformation

The rural sector is declining in importance and its contribution to national economies. It is increasingly part of a unified economy based on the city. Contact with the urban areas is easier than ever and is encouraged by rural development.

Temporary and circular migration is giving way to more permanent settlement. The largest cities are under increasing strain, the residents are encountering increasing difficulties in improving or even maintaining living conditions. Nevertheless, migration continues, driven by a variety of forces both positive and negative. The choice to move can be part of a strategy for survival or personal development; but it is often enforced by external conditions.

The urban transformation is irreversible, but the rural sectors must also be strengthened to balance the developing economy. Attention to gender issue will be crucial in ensuring a successful transition. The forces driving internal and international migration have much in common. Demographic pressures are contributing to both. As the pressures encouraging migration increase, the options for migrants become more limited. This collision is contributing to the atmosphere of crisis surrounding both urban and international migration.

Costs and Benefits

Migration is the result of individual or family decisions. But it is also part of social process. In economic terms, migration is as much a global phenomenon as trade in commodities or manufactured goods. It is part of a broader pattern, and evidence of changing economic, social and cultural relationship.

But migration may be evidence of a different kind of relationship: the combination of poverty, rapid population growth and environmental damage is a powerful destablizing factor driving urban growth and eventually international migration. On the recipient side, migration has usually been seen as evidence of a thriving economy: today's industrial states were built in part by migrant labour, skills and investment. In today's increasingly uncertain conditions, migration may be seen as a threat to the security and well-being of the local workforce and society at large.

The only effective means to reduce migration pressures over the long term are to slow population growth; to stimulate economic growth and job creation at home, and promote the development of the individual and the family as the basic economic and social unit.

A Question of Gender

It is often assumed that most migrants are men, in reality, women make up nearly half of the international migrant population. Gender differences in social and economic roles affect migration decision making, household strategy, and the sex composition of labour migration. Attention to the gender dimension of migratory movements ought to be an important component in population and development planning.

Women frequently take the initiative in migration decisions, which may reflect limited opportunities in rural areas. Low status limits women's choices at home and may increase pressure to migrate, but it may also affect life in the host community. Opportunities may be limited by lack of education or skills, or by customer limitation on women's freedom of action outside the family or ethnic group. Paid employment for migrant women is usually in the lowest wage, least secure, and lowest status jobs, mostly in household, child care and trade.

Most educated women end up in the same low-status, low-wage production and service jobs as unskilled female migrants. Men, too, experience downward mobility, but the contrast in the decline in women's employment status is far greater. Despite these disadvantages women migrants have become significant economic factors. Their status may be improved by migration, but the advantages are not clearcut. Women's status as migrants is affected by their vulnerability, and by their lack of reproductive freedom. To ensure improved status they will need both legal protection and essential services, including reproductive health services.

Refugees

Refugees in the 1990s are overwhelmingly in Asia, Africa and Latin America. Their numbers are large, about 17 million, and growing rapidly. A further 3.5 to 4 million were thought to be in 'refugee-like situations', though estimates are probably extremely conservative, and an estimated 23 million people internally displaced.

It is important to recognize the common roots of refugee and other forms of mass movement of populations. At the same time, despite the difficulty of distinguishing between political and socio-economic causes of migration, there is a clear need to distinguish between refugees and other groups of migrants. Participation in international efforts of burden-sharing would ensure that most refugee problems would be dealt with in their regions of origin.

Conclusions and Policies

Migration highlights linkages and interdependencies with countries, with many implications for development agendas, including population programmes and development assistance.

Policies to regulate or moderate international migration have concentrated largely or urban growth. They have been only intermittently effective. The most successful have concentrated on stimulating rural development and the growth of alternative urban centers.

Migration is also a personal or family decision, which is affected by external conditions such as poverty or environmental degradation, improving conditions of personal and family life can make a crucial difference in the decision to migrate, reducing dependence on migration as a strategy. Because migration is the result of personal and family decisions, it can be influenced by policies that improve the quality of life.

This offers the opportunity for policies emphasizing individual development among them education, health (including reproductive health) and family planning. Such policies are particularly relevant to the strategies must take into account gender differences in social and economic life and the differential effects of policies.

Migration decisions are about family security and long-term-life-chances, rather than simply the maximization of income. They are ultimately strategies designed to look after the individual's and the household's needs, safeguard their security, and respond to their aspirations. If the goal is to reduce migration pressures though development it will be essential to increase the capacity but reduce the need to migrate. Long-term external support will be required to make such policies a reality, particularly in areas of rapid population growth and potential mass outward flows. Highly coordinated allocation of development assistance can help establish priorities and focus attention on basic needs. The challenge to both international donors and co-operating governments is to direct programme spending to the areas where it can be most effective.

Bibliography

Allchin, B. and Allchin, R.,—Civilization: India and The British of Indian Pakistan Before 500 B.C., London, 1968.

All India Economic Conference, *Economic Life of Hyderabad',* Government Central Press, Hyderabad, 1937.

Andhra Pradesh, Government of, *Hand Book of Statistics,* Andhra Pradesh, 1993-94, p. 235.

Andhra Pradesh, Government of *Economic and Statistical Bulletin,* Vol. XXXVII, No. 11, July-December, 1992, Directorate of Economics and Statistics of Hyderabad, p. 10.

Andhra Pradesh, Government of, *Hand Book of Statistics*—Andhra Pradesh, 1993-94, Directorate of Economics and Statistics, Hyderabad, 1995, p. 7.

Andhra Pradesh, Government of, *Statistical Abstract of Andhra Pradesh,* 1992, Directorate of Economics and Statistics, Hyderabad, 1994, p. 111.

Andhra Pradesh, Government of *Statistical Abstract of Andhra Pradesh,* 1992, Directorate of Economics and Statistics, Hyderabed, 1994, p. 111.

Agro-Economic Research Centre, "Rice in Andhra Pradesh—*A Study of Inter State Variations.* Kharif, 1978, Part—I, Report", the Author, Andhra University, Waltair, September, 1982, pp. 15-34 and 59-69.

Apinantara, Adul., "Cooperation and Water Conflicts Among Water Users in North Eastern Thailand Tank Irrigation", ADC, Workshop, Bangkok, Thailand, 1981 (Mimeo).

Appadorai, *A. Economic Research Centre in Southern India* (1000-1500) A.D., University of Madras, 1936.

Arputhraj, C., "*Problems and Prospects of Tank Irrigation in Tamil Nadu*", Workshop on Modernisation of Tank Irrigation, Problems and Issues, Centre for Water Resources, Madras, 1982 (Mimeo).

Chambers, Robert, "*Men and Water: The Organisation and Operation*" in B.H. Farmer (ed), *Green Revolution*, Macmillan, London, 1977.

Cgrabheevyky, P., *Tank Irrigation and Agricultural Development*. Kanishka Publishing House, New Delhi, (India), 1992.

Coward, Sr. E. Walter, *Irrigation and Agricultural Development in Asia*, Cornell University Press, U.S.A., 1980.

Das Gupta, Kalyan Kumar, et al. "Problems of Management of Tank Irrigation", *Impact of Tank Irrigation: A Case Study of Tumkur District*, Himalaya Publishing House, New Delhi, 1978.

Dean. D.J.. and Fray. T.L., "Discriminant Analysis of Loans for Cash Grain Farmers", *Agricultural Economic Review*, Vol. 36 (Annual) (April 1976).

Doherty, Victor S., "*A Cross Cultural Analysis of Tank Irrigation*", Workshop on Modernization of Tank Irrigation: Problems and Issues", Centre for Water Resources, Madras, India, 1982.

Department of Agricultural Engineering, "Development and Optimization in the use of Irrigation Water under Ex-Zamin Tank in Ramanathapuram District", Preliminary Report, Government of Tamil Nadu, Madras, India, 1982.

Elumalai, G., "*Modernisation of Tank Irrigation Systems—Farmer's Views*", Workshop on Modernization of Tank Irrigation: Problems and Issues, Centre for Water Resources, Madras, 1982 (Mimeo).

Evaluation and Applied Research Department, Government of Tamil Nadu, "*Evaluation of Minor Irrigation Schemes*", Second Regional Workshop on Evaluation, Madras, India, 1979.

George. P.T.., Namasivayam. D. and Ramachandraiah. G., "Application of Discriminant Function in the Farmers" Repayment Performance: A Study in Chingelput District, Tamil Nadu", *Journal of Rural Development*, Vol. 3 No. 3, (May 1994).

George W. Snedecor and William G. Cochran, *Statistical Methods*, New Delhi, Oxford and IBH Publishing Co., (Sixth Edition), 1967.

Gupta. S.C., *Development Banking for Rural Development*, New Delhi, Deep and Deep Publications, 1987.

Hand Book of Statistics: Anantapur District, 1991-92 and 1992-93, Chief Planning Office, Anantapur, n.d. p. IV.

Harris, D.G., *Irrigation in India*, Oxford University Press, London, 1923.

Hayami, Y. Bennagem, E. and Barker, R., "Price Incentive Versus Irrigation Investment to Achieve Food Self-Sufficiency in Philippines", *American Journal of Agricultural Economic*, Vol. 59, No. 4, 1977, pp. 717-721.

India, Government of., *Eighth Five Year Plan, 1992-97*, Vol. II, 1992, p. 56.

India, Government of;, "Report on Minor Irrigation Works in the State of Madras" Committee on Plan Projects, 1959.

India, Government of., *Census of India 1991—Final Population Totals,* Vol. II, Series 1, Paper 1, 1992 Ministry of Home Affairs, Government of India, New Delhi, p. 15.

India, Government of, Programme Evaluation Organisation, *A Study of the Problems of Minor Irrigation*, Planning Commission Publication, No. 140, 1961, pp. 7-103.

Indian Agricultural Statistics, for the years 1891-92 to 1947-48.

Janakarajan, S. "In Search of Tanks: Some Hidden Facts", *Economic and Political Weekly*, Vol. XXVII No. 26, June 26, 1993, pp. A-53—A-60.

Jayabalan, B.A., "*Modernisation of Tank Irrigation in Tamil Nadu*", Workshop on Modernisation of Tank Irrigation Problems and Issues, Centre for Water Resources, Madras, 1980.

Jha, U.M., *Importance of Irrigation in an Ararian Economy, Irrigation and Agricultural Development*, Deep and Deep, New Delhi, 1984.

Lakadawala, D.T., "*Growth, Employment and Poverty*", Presidential Address, All India Labour Economics Conference, Tirupati, December, 1977.

Madras Institute of Development Studies, "*Tank Irrigation in Tamil Nadu—Some Macro and Micro Perspectives*", (Report—Mimeo), 1983.

Mukundan, T.M., "*The ERY System of South India*, "PPST Bulletin, Madras, September, 1988.

Namerta, "*Growth and Spatial Pattern of Market Towns in Rayalaseema Region of Andhra Pradesh*" Unpublished M.Phil Thesis, submitted to Jawaharlal Nehru University (Centre for Development Studies), New Delhi, 1989.

Narayana Rao, J.S., "*Irrigation and Planning in Andhra Pradesh*", Conference Papers of Andhra Pradesh Economic Association, Sixth Annual Conference, 23-24 January, 1988, pp. 145-150.

National Institute of Rural Development, *Rural Development Statistics: 1994*.

Narain, Dharan, and Roy B. Shyamal, "*Impact of Irrigation on Labour Availability and Multiple Cropping*", International Food Policy Research Institute, Research Report, 2-0, 1980, pp. 7-8 and 26.

"Need for Restoration of Tanks:, *The Hindu*, August 8th, 1991, p. 3.

Narain, et al., "An Approach to Study of Irrigation—A Case Study of Kanyakumari District:, *Economic and Political Weekly*, Vol. XVII, No. 39, September 25, 1982, p. 485.

Palani Swamy, K., "Irrigation Tank Rehabilitation", *The New Irrigation Era*, Vol. XIX, No. 3, 1981, pp. 36-37.

Palaniswamy, K. and Easter, K.W., "*The Tanks of South India: A Potential for Future Expansion in Irrigation)*", Economic Report, Department of Agriculture and Applied Economics, University of Minnisote, No. ER 83.4, 1983.

Rao. G.N. and Rajasekhar, D., "*Tank Irrigation in Andhra Pradesh: A Case Study*" Conference Papers of Andhra Pradesh Economic Association, Second Annual Conference, January, 1985, pp. 57-70.

Rao, V.M., "Linking Irrigation with Development—Some Policy Issues", *Economic and Political Weekly*, Vol. XII, No. 24, 1978, pp. 993-97.

Rajasekhar, D., *Land Transfer and Family Partitioning* Oxford and IBH Publishing Company, New Delhi and Centre for Development Studies. Trivandrum, 1988.

Rao, N.V.N., and Ram Reddy, R., "Minor Irrigation and Tribal Development—An Empirical Study", *Kurukshetra*, Vol. XXXIV, No. 2, November 1995, pp. 31-33.

Ramaswamy, V., et. al., "Damasi: A Concept of Equity and Productivity in Irrigation", *Wamana*, Vol. V, No. 3, July, 1985, pp. 1 and 15-22.

Ramanathan, "*Tank Irrigation in Tamil Nadu*", Unpublished M.Phil Thesis, The Jawaharlal Nehru University, 1985, pp. 70-137.

Reddy, Narasimha, D., "A Note on the Decline of Tank Irrigation", Paper Presented at Lokayan Workshop on Tank Irrigation, Bangalore, July, 1988.

Rao, V.M., "Linking Irrigation with Development—Some Policy Issues", *Economic and Political Weekly*, Vol. XII, No. 24, July 17, 1978, pp. 993-997.

Rami Reddy. S., "Factors Discriminating Defaulters from Non-Defaulters in Primary Credit Cooperatives", *Indian Cooperative Review*, Vol. 14, No. 1 (October 1976).

Satis, S. and Sundar, A., *People's Participation and Irrigation Management: Experiences, Issues and Options*, Commonwealth Publishers, New Delhi, 1990.

Sakthivadivel, R., et. al., "*A Pilot Project Study of Modernization of Tank Irrigation: Problems and Issues*", Centre for Water Resources, Madras, 1982 (Mimeo).

Sen Gupta, Nirmal., *Managing /Common Property: Irrigation in India and the Philippines*, Sage Publications, New Delhi, 1991.

Sen Gupta, Nirmal, "*Irrigation: Traditional Vs. Modern*", Working Paper No. 55, Madras Institute of Development Studies, Madras, April 1985, p. 5 (Mimeo).

Sen Gupta Nirmal, *Tank Irrigation in Gangetic Bihar*, A.N.S. Institute of Social Studies, Patna, Bihar, 1982.

Sivanappan, R.K., "*Water Management in Tank Irrigation in Tamil Nadu*", Workshop on Modernisation of Tank Irrigation: Problems and Issues, Centre for Water Resources, Madras, 1982 (Mimeo).

Sivamohan, M.V.K., and Christopher A. Scott, (Ed), *India: Irrigation Management Partnerships*, Booklinks Corporation, Hyderabad, 1994.

SPSS Inc., SPSS/PC Release 1, 1 Update, the Author, U.S.A., 1984, pp. B. 202-204.

Sundar, A. and Rao, P.S., "*Farmers' Participation in Tank Irrigation in Karnataka*", Workshop on Modernisation of Tank Irrigation "Problems and Issues, Centre for Water Resources, Madras, India 1982.

Srinivasan, T.M., *Irrigation and Water Supply: South India 200 B.C., 1600 A.D.*, New Era Publications, Madras, 1991.

Subbalkshmi, V., "*Incorporation of India and Indigenous Irrigation Institutions: The Case of Dasabandam in Rayalaseema*", Conference papers of Andhra Pradesh Economic Association, Sixth Annual Conference, January, 1988, pp. 114-128.

Smith R. Baird, *Irrigation in Southern India*, Elder and Co., London, 18567, Ch. 6.

Tubpin, Y., Easter, K.W;, Welsch, D., "*Tank Irrigation in North-Eastern Thailand, The Returns and Their Distribution*", Economic Report,

Department of Agricultural and Applied Economics, University of Minnesota, No. Er, 82-6, 1982, p. 66.

Uma Shankari, "*Tanks: Major Problems in Minor Irrigation*", *Economic and Political Weekly*, Vol. XXVI, No. 39, September 28, 1991, pp. A115-A125.

USAID, "*Thailand North-East Small Scale Irrigation*", Washington, D.C., 1980.

Vasudeva Rao, D. *Rural Development Through Irrigation*, New Delhi, Ashish Publishing House, 1987.

Venkatram, B.R., "*Administrative Feasibility of Tank Irrigation Authority*", Economics Programme, Consultancy Report, ICRISAT, Hyderabad, Andhra Pradesh, 1980, pp. 1-4.

Von Oppen, M. and Subba Rao, K.V., "*Tank Irrigation in Semi-Arid Tropical India, Part—I: Historical Development and Spatial Distribution*", ICRISAT, Economics Programme Progress Report, 5, Andhra Pradesh, India, 1980, p. 1.

Von Oppen, M. and Subba Rao K.V., "*Tank Irrigation in Semi-Arid Tropical India, Parts I, II, III*", Progress Report, ICRISAT, Hyderabad, 1980.

Von Oppen, M. and Subba Rao, K.V., "*History and Economics of Tank Irrigation in Semi-Aluid Tropical India*" Symposium on Rain Water and Dry Land Agriculture, Indian National Science Academy, New Delhi, 1982, pp. 89-93.

Von Oppen, M., "*Tank Irrigation in Southern India: Adopting a Traditional Technology to Modern Socio-Economic Conditions*", Proceedings of the Consultants' Workshop on State of the Art and Management Alternatives for Optimizing the Productivity of SAT Alfisoils and Related Soils, ICRISAT, Patacheru, Hyderabad, Andhra Pradesh, India, 1987, pp. 89-93.

Wijayaratna, C.M., "*Uneven Distribution of Water: Causes, Consequences and Implications for System Management*" Workshop on Water Management, Agrarian Research and Training Institute, Colombo, Sri Lanka, 1982, (Mimeo).

Yazdani, G. (Ed), *The Early History of Deccan*, Oxford University Press, London, Parts—VI, 1960.

Index